AF522289

Scientific Horticulture

Scientific Horticulture

Abhishek Guneta

RANDOM PUBLICATIONS
NEW DELHI (INDIA)

Scientific Horticulture

ISBN 978-93-5111-367-6

Published in 2014 in India by

RANDOM PUBLICATIONS

4376-A/4B, Gali Murari Lal, Ansari Road
New Delhi-110 002
Phone : +91-11-43580356, +91-11-23289044
e-mail: randomexports@gmail.com, sales@randompublications.com,
info@randompublications.com

Reprint 2021

Type Setting by: Friends Media, Delhi-110089
Digitally Printed at: Replika Press Pvt. Ltd.

Preface

Horticulturists apply the knowledge, skills, and technologies used to grow intensively produced plants for human food and non-food uses and for personal or social needs. Their work involves plant propagation and cultivation with the aim of improving plant growth, yields, quality, nutritional value, and resistance to insects, diseases, and environmental stresses. They work as gardeners, growers, therapists, designers, and technical advisors in the food and non-food sectors of horticulture. Horticultural scientists focus on the research that underpins horticultural knowledge, skills, technologies, education, and commerce. Horticultural science encompasses all of the pure sciences – mathematics, physics, chemistry, geology, and biology – as well as related sciences and technologies that underpin horticulture, such as plant pathology, soil science, entomology, weed science, and many other scientific disciplines. It also includes the social sciences, such as education, commerce, marketing, healthcare and therapies that enhance horticulture's contribution to society. A gardener is a person that tends to a garden and is therefore a horticulturist. However, not all horticulturists are gardeners.

Horticulture is a term that evokes images of plants, gardening, and people working in the horticultural industries. For the public, and policy makers, the term is not completely understood nor is its impact on human activities been fully appreciated. Horticulture impacts widely on human activities, more than its popular understanding as merely "gardening" would indicate. It needs to be recognised as a matrix of interrelating areas that overlap, with complex interrelationships. A wider and more accurate definition will communicate effectively the importance of plants, their cultivation and their use for sustainable human existence. The popular "gardening activity" sense fails to convey the important role that horticulture plays in the lives of individuals, communities and human societies as a whole. Describing its impact on the physiological, psychological and social activities of people is key to expanding our understanding; however "the cultivation of a garden, orchard, or nursery" and "the cultivation of flowers, fruits, vegetables, or ornamental plants" as well as "the science and art of cultivating such plants" will suffice to sketch the outline of a short description. Relf (1992) expanded the traditional understanding of horticulture beyond "garden" cultivation. Tukey (1962) gave an overview of those involved in the field of horticulture, in stating that there are those who are concerned with the science or biological side, those concerned with the business side and finally those who are concerned with the home or art side, who enjoy plants simply for the satisfaction they get from them. Primarily it is an art, but it is intimately connected with science at every point. Relf highlighted the fact that limiting the definition of horticulture severely limits an understanding of what horticulture means in terms of human well-being. Relf provided a comprehensive definition of horticulture as; the art and science of plants resulting in the development of minds and emotions of individuals, the enrichment and health of communities, and the integration of the "garden" in the breadth of modern civilisation. In addition, Halfacre and Barden (1979), Janick and Goldman (2003). Further extended the scope of horticulture when they agreed that the origins of horticulture are intimately associated with the history of humanity and that horticulture encompasses all life and bridges the gap between science, art and human beings. This broader vision of horticulture embraces plants, including the multitude of products and activities (oxygen, food, medicine, clothing, shelter, celebration or remembrance) essential for human survival; and people, whose active and passive involvement with "the garden" brings about benefits to them as individuals and to the communities and cultures they encompass.

This book represents an essential source of up-to-date practical information on this subject.

I thank all members of my team who have helped in the preparation of the book. My special thanks go to "Random Publications" who have published the book.

—Abhishek Guneta

Contents

1

Conservation and Utilisation of Horticultural Genetic Resources

This paper has been developed on the basis of existing FAO documentation and, in some instances, by collating information extracted from documents prepared during the last 10 years for the FAO Commission on Genetic Resources and various other FAO technical meetings related to the conservation and utilisation of plant genetic resources for food and agriculture. Although some of the considerations developed in this paper might seem to address issues of a general character, they are nonetheless quite applicable to horticultural genetic resources. The document should be regarded as an approach intended to stimulate reflection during the elaboration of a global research policy for the development and improvement of horticultural production through the conservation and best use of horticultural genetic diversity.

Bio-diversity and Food

Bio-diversity is a popular way of describing the diversity of life on earth: it includes all life forms and the ecosystems of which they are a part. Bio-diversity forms the foundation for sustainable development. It is the basis for the environmental health of our planet and a source of economic and ecological security for future generations. In the deveic and ecological security for future generations. In the developing world, bio-diversity provides the assurance of food, countless raw materials such as fibre for clothing, materials for shelter, fertilisers, fuel and medicines, as well as source of work energy in the form of animal traction.

The rural poor depend upon biological resources for an estimated 90% of their needs. In the industrialised world, access to diverse biological resources is necessary to support a vast array of industrial products. In the continuing drive to develop efficient and sustainable agriculture for many different conditions, these resources provide raw material for plant and animal breeding, as well as for the new biotechnologies. Genetic diversity in agriculture enables crops and animals to adapt to different environments and growing conditions. The ability of a particular variety to withstand drought or inundation, grow in poor or rich soil, resist one of the many insect pests or diseases, give higher protein yields or produce a better-tasting food are traits passed on naturally by its genes. This genetic material constitutes the raw material that plant and animal breeders and biotechnologists use to produce new varieties and breeds. Without this diversity we would lose the ability to adapt to ever-changing needs and conditions. Sustainable agriculture could not then be achieved in many of the world's different food production environments.

Different Food Production Environments

Diversity among individual plants and animals, species and ecosystems provides the raw material that enables human communities to adapt to change - now and in the future. Deprived of bio-diversity, the ability of humankind to meet the challenges resulting, for example, from climate change would be severely limited. The diversity found within the small number of plant and animal species which form the basis of world agriculture and food production remains a small but vital part of the earth's bio-diversity.

Genetic diversity is not evenly distributed throughout the world; it is in fact concentrated in tropical and sub-tropical areas, where the majority of developing countries are located. Plant genetic resources are of an inestimable value, and will continue to be so in the future, independently of whether scientists use them by means of conventional plant breeding or modern genetic engineering. The importance of plant genetic resources as the ultimate source of food is enormous. Their loss constitutes a serious threat to world food security. The conservation of genetic resources goes far beyond the salvation of species. The objective must be to conserve sufficient diversity within each species to ensure that its genetic potential will be fully available in the future.

Within the Plant Kingdom, around 350,000 species have been classified, out of which about 80,000 have been classified, out of which about 80,000 have been found to be edible. In the course of history, mankind has utilised about 7,000 of these plant species for food.

Today, only 150 plant species are cultivated and of these, the so-called major crops can be contemplated in about 30 plant species which are producing about 95% of the world's calories and proteins. About 75% of food consumption comes from only 12 plant species and 5 animal species. Half of this food comes from only 4 plant species (rice, maize, wheat and potato) and 3 major animal species (cattle, swine and poultry).

Over millennia, men and women farmers have developed within each domesticated food species thousands of landraces (farmers varieties) and breeds adapted to local conditions and needs. This "human-made" agro-biodiversity is now seriously endangered. Today the world population amounts to about 5 billion people. An increase of about 60% is expected in the next thirty years, which will bring the world population to some 8 billion people. Today, 800 million people are chronically malnourished. On the basis of recent estimates, in the developing countries alone, food production will need to increase by about 60% in the next 25-30 years in order to keep pace with the expected demographic growth.

Losing Plant Bio-diversity

Bio-diversity maintains the ecological balance necessary for planetary and human ecological balance necessary for planetary and human survival. Biological resources are renewable resources, but they are being exploited at rates that exceed their sustainable yield. Human destruction of habitats, whether exploited for commercial or subsistence reasons, is the greatest threat.

The clearing of land for agriculture, overgrazing of grasslands, cutting and burning of forests, unsustainable logging and fuel wood collection, indiscriminate use of fertilisers and pesticides, over-watering of crops, over-exploitation of fisheries, draining and filling of wetlands, poor water management, urbanisation and pollution of air and water, all figure prominently as causes of the degradation of our biological resources.

Genetic erosion is a term often used to describe the reduction of diversity within a species and, as the main cause of species extinction, it is a global threat to agriculture. If the individuals we conserve from any one species contain only a small fraction of the total species gene pool, then their ability to adapt to changing circumstances in the future will be severely limited. Genetic erosion is usually the loss of genetic diversity as a result of social, economic agricultural changes. UNCED's Agenda 21 states that *'the current decline in bio-diversity is largely the result human activity and represents a serious threat to human development'.* This statement is also applicable to erosion of genetic ment is also applicable to erosion of genetic resources for food and agriculture.

In modern agriculture, food habits, and hence crops, have often been introduced and spread by dominant cultures without necessarily taking into account local environmental conditions, or the needs of local communities. The introduction of bread in many parts of the world, for example, saw wheat overtake many traditional local staple crops, which were often well suited in the local environment and highly productive.

The alarming rate of global bio-diversity losses must be of concern to everyone. The empirical and potential estimates for species extinction caused by habitat destruction and pollution suggested by Lugo (1988) range from one species/day to one species/hour between 1970 and 2000. By the end of this century, our planet could have lost anywhere from 20 to 50 percent of its species or 500,000 to several million, including 35,000 to 40,000 plant species. Until the quite recent advent of scientific plant breeding, all of the crops that farmers grew were "landraces"; that is, local crop varieties developed in traditional agricultural systems with selection - both natural and by farmers - over long periods of time. In the late 19th century, plant breeders using the emerging science of genetics began to systematically to improve the land-races. From this process emerged a whole series of more advanced cultivars emerged a whole series of more advanced cultivars or modern varieties. This process has led to steady increases in global yields in this century, allowing food production to follow the trend of the world's growing population. However, a product of plant breeding has been that it has tended to narrow the genetic base of a crop in rough proportion to its success.

In all advanced technology-based agriculture, a few excellent varieties, themselves often genetically inter-related, now tend to cover large areas of land to the exclusion of all else. The side-effect of this widespread planting of a relatively small number of improved varieties with special genetic characteristics has been widespread loss of the varieties which preceded them.

There are numerous examples among horticultural crops of genera and species which have suffered or are vulnerable to the threat of genetic erosion. The genus *Malus* is probably the case most often cited in this respect, with the world's apple industry being now largely based on two varieties and a few of their progeny, and with an estimated 85% of the 7,000 cultivars formerly known and described having been lost.

For many people this displacement of traditional varieties represents a process of genetic erosion at the farm level where not only individual genes are lost but also complexes which are specifically adapted to particular local environments. There is now a pted to particular local environments. There is now general agreement among scientists that the conservation of the remaining crop genetic resources is urgent and essential to maintaining long-term food security.

Genetic erosion, or the loss of bio-diversity, is often presented as an ecological problem, but the underlying fundamental causes are socio-economic and political. The lack of economic incentives to conserve diversity within agriculture and nature is a key cause of continuing degradation. It is a vital challenge for our societies to properly reflect the precious value of nature's diversity in the world economy, in all market exchange rules and regulations. Currently the basic genetic raw material with which breeding and biotechnology industries work is essentially available free of charge from nature or farmers' fields.

The cost of conserving bio-diversity is high, but almost certainly far less than the cost of allowing its degradation. While modern biological science and technology can do wonders, it has to be remembered that species extinction is forever. The world therefore needs a means of valuation which recognises that the loss of biological resources is irreversible. Our generation has a tremendous social responsibility: to pass on to our children the integrity and wealth of Bio-diversity that we have inherited from our parents. Only by doing this will we enable future generations to face unpredictable enable future generations to face unpredictable environmental changes and human needs.

The Use of Biotechnology

Through modern biotechnologies that are helping to overcome some of the limitations of conventional breeding, wild diversity can be more readily incorporated into crops and thus contribute to world agricultural development.

It is recognised that production must be intensified, productivity increased and productive natural systems must be suitably managed - all in a sustainable manner. This requires the combined application of new and old technologies, including innovative approaches to plant breeding and to farming practices. The success of such endeavours will depend on the sustainable utilisation of a broader range of species and genetic material within each species, including the wild relatives of domestic species.

Many of the recent advances in agricultural technology, including biotechnologies, have been made in industrialised countries where they have often been developed on a commercial basis. In this sense, the potential for the new biotechnologies to improve the livelihood and food security situation of the world's poor is largely under-realised. The new biotechnologies do provide excellent tools for manipulating the genetic diversity of crop varieties and animal breeds among the already domesticated species and further increasing productivity and stabilspecies and further increasing productivity and stability. Through the development of plants and animals that are more adapted to ecological conditions, the stability of production can be increased. The genetic manipulation of other biological elements in crop production systems, such as soil bacteria associated with nitrogen fixation, or natural predators of pests (especially insects and to some degree pathogens and weeds) is also a tool that can be used for enhancing sustainability in agriculture.

Ours is the first generation with the immense biotechnological power to radically change the evolutionary processes of natural and artificial selection. Ours, too, is the responsibility that goes with this power.

However it is recognised that there are also risks involved in the application of biotechnologies; these risks include misuse by humans, and environmental accidents. To address these and other issues, a draft Protocol on Bio-safety is now being negotiated by those countries which are signatories to the Convention on Biological Diversity. With the advent of genetic engineering, the private sector biotechnology industry has promoted the extension of industrial patenting regimes to living organisms - an approach popularly known as life patenting.

As a result, genes, micro-organisms and varieties of higher species (plants and animals), either discovered in nature or genetically manipulated by er discovered in nature or genetically manipulated by breeders or genetic engineers, can and have become the intellectual property of individuals or industry in countries in which "life patenting" has been legalised. Between 1981 and 1985 at least 1175 patents for human genes were granted worldwide; and of these patent applications, some 75% were from the private sector.

Perhaps for reasons guided by the market, the socio-economic and food security needs of the poorer sections of society, especially in developing countries, have not yet been sufficiently targeted by biotechnology programmes, or have not been adequately considered in shaping research policy.

This is an important point for decision-making in relation to the allocation of public sector funding for agricultural research, since it is estimated that there are now at least 1,400 million people in developing countries who are dependent upon resource-poor farming systems which have not benefited from technological advances in plant or animal breeding or biotechnology.

It can also be debated whether current intellectual property right systems allow appropriate incentives for directing public sector agricultural research so that the social and economic needs of these poor countries can be adequately met. This issue becomes particularly worrying when it is considered that the limitation of access for the world's poor farmers to improved genetics for the world's poor farmers to improved genetic materials could become a constraint to increased food production.

The FAO Global System

Within the programmes of the Food and Agriculture Organisation of the United Nations (FAO), the inter-related questions of who benefits from the use of genetic resources and who owns genetic resources stored in international seedbanks, in farmers' fields, or in nature, became the focal point of international debates during the 1980s. To confront and address these problems, the member countries of FAO agreed on the need to develop a global system which might ensure the conservation and sustainable use of plant genetic resources for food and agriculture, as well as the just and equitable sharing of the benefits and responsibilities derived from them.

The development of the Global System on Plant Genetic Resources for Food and Agriculture began in 1983 with the establishment of its governing body, the Commission on Plant Genetic Resources, now the Commission on Genetic Resources for Food and Agriculture, a permanent inter-governmental forum for discussion and consensus building between countries. Governments also adopted, that same year, an international non-binding agreement, the first of its kind, called the International Undertaking on Plant Genetic Resources, which provides the formal frameant Genetic Resources, which provides the formal framework for the Global System. The Commission has since coordinated, overseen and monitored the development of a Global System for the Conservation and Utilisation of Plant Genetic Resources for Food and Agriculture.

Countries continue to negotiate and develop, through the Commission, other important elements of the System which include various international agreements, scientific regulations, technical cooperation mechanisms and global instruments which are in different stages of development.

The availability of information about plant genetic resources of crop species is fundamental to being able to harness the genetic variability conserved in collections and critical for evaluating the usefulness of *in-situ* conservation initiatives to counter the risk of genetic erosion. In response to this need, the FAO World Information and Early Warning System on Plant Genetic Resources (WIEWS) was created five years ago, within the overall framework of the FAO Global System on Plant Genetic Resources. Its primary purpose is to facilitate access to information so as to allow a more coherent utilisation of available genetic resources for plant improvement work.

Plant Genetic Resources

Despite the useful progress made so far, some major questions are still currently being debated in the FAO fora that deal with plant genetic resources for food and agriculture, *i.e.*:

- Who owns plant genetic resources?
- Whose responsibility is it to conserve them?
- Who benefits from their use?
- What form the benefits deriving from their use should take, and how should they be shared?
- How can the benefits of biotechnology be directed to the largest number of people and countries?
- How can measures be taken to prevent certain applications of biotechnology endangering the livelihoods of large numbers of people in developing countries? (for example by new products replacing the agricultural commodities that provide their livelihood)?
- How can biotechnological research be directed towards the maximum benefit to world society?
- What measures can be taken to prevent modern, intensive agriculture, with homogeneous varieties, stamping out the

world's rich inheritance of plant germplasm conserved in the heterogeneous local varieties in farmers' fields?

In June 1996, the Fourth International Technical Conference on Plant Genetic Resources, which was attended by 150 countries and 54 inter-governmental and non-governmental organisations, was held in Leipzig, Germany. The Conference was convened at the request of the FAO Commission on Plant Genetic Rthe request of the FAO Commission on Plant Genetic Resources, and was endorsed in Agenda 21 of the United Nations Conference on Environment and Development, and at the Nairobi Conference for the Adoption of the Agreed Text of the Convention on Biological Diversity.

The Conference adopted a Global Plan of Action for the Conservation and Sustainable Utilisation of Plant Genetic Resources, together with the Leipzig Declaration, and it also considered the first Report on the State of the World's Plant Genetic Resources.

The Global Plan of Action was prepared through a participatory, country-driven process, involving a wide variety of stake-holders: governments, non-governmental and industry organisations, and individual scientists. A total of 158 governments prepared Country Reports, assessing the status of their plant genetic resources, as well as their capacity to care for and utilise these resources. Twelve regional and sub-regional meetings were held, where governments considered regional problems and opportunities, and made recommendations for the Plan. The Report on the State of the World's Plant Genetic Resources is the first comprehensive world-wide assessment of the state of plant genetic resource conservation and use, identifying the urgent priorities for action which are, in turn, addressed in the Global Plan of Action. The Report and the Plan are now two strategic elements of the Plan are now two strategic elements of the FAO Global System for the Conservation and Utilisation of Plant Genetic Resources.

The Plan is intended to provide a coherent strategy for action in several fields: *in-situ* and *ex-situ* conservation, sustainable utilisation of plant genetic resources, and institutional and capacity building. Therefore, the GPA needs to be taken as a whole and considered in a comprehensive manner embracing all operational, programmatic, policy and political aspects.

The Leipzig Declaration asserts that "our primary objective must be to enhance world food security through conserving and sustainably using plant genetic resources". This Declaration commits the

Governments to taking necessary steps to implement the Global Plan of Action.

On a different, albeit related, plane, the world's political leaders, meeting in Rome at the 1996 World Food Summit, have made a public commitment to end hunger. Through the Plan of Action adopted by the Summit, governments, international organisations and all sectors of civil society are encouraged to join forces in a concerted effort to ensure access at all time to the food required for a healthy active life for all the world's people. Meeting the same month, in November 1996, the Third Conference of the Parties of the Convention on Biological Diversity held in Buenos Aires focused on agricultural bio-diversity, and its res focused on agricultural bio-diversity, and its role in ensuring a transition towards sustainable agricultural practices.

All of these inter-governmental meetings recognised the importance of plant genetic resources for food security and called on countries to implement the Global Plan of Action for the Conservation and Sustainable Use of Plant Genetic Resources for Food and Agriculture.

The needs highlighted at Leipzig, Rome and Buenos Aires, and the recommendations agreed upon by governments in the Global Plan of Action for Plant Genetic Resources present many challenges to all crop scientists, in terms of developing or adapting suitable technologies and approaches for the sustainable utilisation of plant genetic resources for food and agriculture (PGRFA). The concept of sustainable utilisation is defined in the Convention on Biological Diversity as the use of genetic resources in a productive manner which does not lead to the long term loss of such resources. This provides for an approach which integrates both conservation and utilisation and which spans all aspects of the use of PGRFA, whether by scientists or farmers.

Plant genetic resources can be described as the part of bio-diversity that nurtures people and is nurtured by people. Horticultural crops, although constituting only one part of plant genetic resources for food and agriculture, comprise a very wide range of species of vital importance to ensuring age of species of vital importance to ensuring food security.

All governments have now recognised that the utilisation of plant genetic resources is the key to improving agricultural productivity and sustainability and can contribute to socio-economic development, food security and the alleviation of poverty. To achieve these desirable goals, the Global Plan of Action promotes an integrated strategy for the conservation and sustainable utilisation of PGRFA with the following features: *productivity* - greater use of plant genetic resources will be required to contribute to the productivity increases needed to meet growing populations. This will require continued access to, and exchange of, the world's plant genetic resources; *sustainability* - there is a need to ensure that such use of plant genetic resources is coupled with the conservation of plant genetic resources, both *in-situ* and *ex-situ*.

This will require, *inter alia*, approaches to crop improvement which allow the maintenance of higher levels of genetic diversity and resources in production systems, thereby contributing to reduced genetic vulnerability and less genetic erosion. Additionally, conservation programmes should be clearly linked with utilisation efforts and the sharing of benefits, in order to reinforce the sustainability of such programmes; *equity* - those responsible for conserving and developing plan those responsible for conserving and developing plant genetic resources should be able to participate fully in the benefits derived from their use. There is thus a need to develop crop improvement approaches to enable farmers in marginal areas, as well as those in high-productivity areas to benefit fully from the utilisation of plant genetic resources.

In meeting these challenges of harnessing plant genetic resources for sustainable agriculture and future food security, the role of crop scientists will be crucial.

Global Networks

The importance of conserving plant genetic resources has been stressed in several international agreements, such as the International Undertaking on Plant Genetic Resources and the Convention on Biological Diversity and its Agenda 21. The successful conservation and sustainable utilisation of plant genetic resources for food and agriculture involves action by a wide range of people in every country: policy makers, planners, scientists, genebank curators, breeders, rural communities and farmers. Strong coordination mechanisms are required at the national level to enable all these players to participate constructively.

In response to this need, through its international networks on *ex-situ* and *in-situ* conservation, FAO is pursuing efforts in order to coordinate, under this umbrella, the conservation and utilisation of er this umbrella, the conservation and utilisation of plant genetic resources of many agricultural crops and to facilitate the exchange of genetic material, information and technology.

The need for germplasm conservation is recognised for all crops and plants of agricultural importance. This, in turn, dictates the need for programmes in evaluation and documentation, and these activities must be carried out in cooperation with institutions, scientists and technicians with crop-specific competence. In this context, we consider it important to foster the establishment of crop-related networks with special competence and responsibilities related to genetic resources of specific crops.

The development of clearer policy lines related to the conservation of genetic variability of crop species, through regional and global inter-country cooperative programmes, was one of the principal aims of the Leipzig Technical Conference on Plant Genetic Resources. The Global Plan of Action on PGR is now under implementation through a partnership mechanism involving inter-country crop-related networks, as well as several national and international institutions and organisations, including non-governmental organisations. The Global Networks which have been recently launched by the FAO are thus elements of the Global Plan of Action.

It is being demonstrated that inter-country networking constitutes a useful instrument to facilitate policy itutes a useful instrument to facilitate policy development and to provide guidance and global coordination for inter-institutional research work concerning crop genetic resources. It should be underlined that the crop-related networks that operate under the aegis of the FAO have been constituted on the basis of voluntary and self-funding participation of interested countries, institutions and scientists.

In facilitating the implementation of the Global Plan of Action on PGR, and recognising the important role of horticultural crops for agriculture and for food security, FAO has concentrated efforts on promoting regional, inter-regional and global crop-related networks, particularly oriented towards the horticultural sector, as instruments to facilitate scientific exchange, information sharing, technology transfer, and research collaboration. The networks also provide a forum for determining ways of sharing responsibility for the collection, conservation, evaluation, characterisation and utilisation of horticultural genetic resources. These crop-related networks bring together different types of specialists to set collaborative research objectives, define policy priorities and strengthen activities on the conservation and utilisation of genetic resources in relation to specific groups of crops. Most of the partners of these inter-country cooperative programmes are internationally recognised scientists and members of the Internatignised scientists and members of the International Society for Horticultural Science which further strengthens collaboration between the ISHS and FAO.

The innovative aspect of these networks is that they promote a co-ordinated approach to identifying, evaluating and conserving the genetic variability of selected crop species, with the aim of improving cultivars and adapting them to farmers' needs. The networks combine a thorough knowledge of the agricultural conditions of farmers in network member countries, with an understanding of the genetic potential of the crop species in question, and use a "farmers-to-farmers" basis for the development of conservation and utilisation activities. That involves the collection from farmers of locally adapted germplasm, its improvement and its return to the same farmers, or farmers in similar biotopes. The work of several of these networks is also guided by an economic intelligence function (analysis of market factors), which helps farmers to improve quality and suit their production to market requirements.

The distinguishing characteristic of the networks' components on plant genetic resources consists in identifying the genetic variability within each particular crop species, and the development of methodologies for characterisation, evaluation and conservation of different genotypes in view of their utilisation by present and future generw of their utilisation by present and future generations of farmers. As such, the networks are regarded as very useful instruments for facilitating the process of implementing the Global Plan of Action, and their expansion, in terms of both crops and regions covered, is being pursued.

The crop-related networks also provide a useful forum for updating the Report on the State of the World's Plant Genetic Resources. Computerised data bases are being developed for several of the networks and the information being gathered is available to be systematically included in the World Information and Early Warning System.

The many crop-specific networks and working groups may operate at sub-regional, regional, inter-regional or global level. The networks that FAO has developed over the last seven years, with the aim of promoting a coordinated approach to identifying, evaluating and conserving the genetic variability of selected crop species, include the Global Mushroom Germplasm Conservation Network; the Olive Genetic Variability Conservation Network; the International Network on Cactus Pear; the Mediterranean and Inter-American Citrus Networks and the Global Citrus Germplasm Network; the Inter-regional Cooperative Network on Nuts; the Global Network on Tropical and Subtropical Fruit Genetic Resources, and the Network on Traditional Crops of Southern Africa.

The International Mushroom Germplasm Conservation

Fungi are regarded as being the second largest group of organisms in the biosphere after the arthropods. The total number of fungal species is evaluated as 1,500,000 in the world. Only 5% of theses species are described and catalogued. Therefore our knowledge on the worldwide fungal species is poor. Out of the 70% described species of fungi, there are about 10,000 species of mushrooms in which about 2,000 species from more than 30 genera are regarded as prime edible mushrooms, but only a few dozen of them are cultivated. In the particular of tropical and sub-tropical areas, the fungal and mushroom genetic resources are certainly the most abundant of the world and the least known.

This network was established to strengthen international collaboration among specialised institutions, with a view to constituting a coordinated global system of mushroom germplasm collections under the aegis of FAO, and to facilitate technical communication and the exchange of strains of cultivated mushroom, and of germplasm of other mushroom species of interest for food and agriculture. The network also has the aim of establishing a more comprehensive and coordinated information mechanism, which would help to make mushroom strains, together with appropriate production technologies, available to as many interested countries and growers as possible.

2

Global Horticultural Impact

Since 1985 horticultural produce and commodities have been evincing increasing importance in developing countries, which have increased their overall share of world production and now have fruits and vegetables making up a large proportion of their exports. Future growth will likely depend on rising and changes in demand in both developed and developing countries and on the capability of emerging economies to maintain or increase their competitive strength in world markets. The implementation of the Uruguay Agreement will impact on trade volumes and trade flows particularly through such general areas as market access, tarification, limits on export subsidies, cuts in domestic supports, phytosanitary measures and safeguard clauses. The brief of the present study is to gain a better understanding of the economic significance of, and outlook for, production, consumption, imports and exports of fruits and vegetables, and key issues related thereto, in developing countries.

In recent years analysts have been taking a much closer look at the horticultural sector of developing countries, especially in regard to its potential contribution to income generation and export earnings (OECD, 1996; Islam, 1990). The literature is full of specific studies on selected horticultural produce/countries (Segre *et al.*, 1997), although little research has been done to arrive at a comprehensive analysis of production, consumption and trade. One factor explaining the lack of such a comparative approach is the limited range of available data.

On the strength of a homogeneous FAO-derived worldwide database (World Tables of Fruits and Vegetables Development, WTF&VD) specifically worked up for its brief, the present paper adds to this small fund of knowledge so that a better understanding of the economic significance of and outlook for production, consumption, imports and exports of fruits and vegetables (excluding such products as cut flowers, foliage, potted plants, and nursery materials) in developing countries can be achieved. The data reported in the following pages cover 61 products over the period 1985-1996 for 168 countries. The following sections also summarize certain findings that were part of a wider analysis conducted in the framework of the WCHR-Working Group 1. The complete information set, including a widely cast bibliographic net, can be found a widely cast bibliographic net, can be found in the working document by Segre et al. (1997), while the summary tables of the database are now on the WCHR Web Site (Segre et al., 1998) and the country-by-country and product-by-product dataset will be available upon request to the author.

Methodology

The present study combines statistical and economic analyses. The former is performed by the WTF&VD. These tables, derived from the FAO database, cover the time period 1985-1996 and consider some 228 countries arranged in 17 country groupings. Only the aggregate "Developing Countries" (hereinafter DCs) is considered here, which enumerates nine country groupings: 1. North Africa, 2. Sub-Saharan Africa, 3. Central America and Caribbean, 4. South America, 5. China, 6. Near East in Asia, 7. East and South-East Asia, 8. South Asia, 9. Oceania Developing. Within each one three product groups are examined-vegetables, fresh fruits and other fruits-and for each such group in turn the following indicators are reported: production, imports and exports (in volume and value). Five-year averages, annual growth rates and shares were than calculated, and total and per capita (apparent) consumption and a standardized trade balance index were derived. It should be noted that in the WTF&VD all these figures are available also at comp; VD all these figures are available also at country and product levels.

These data were then used to apply an economic model known in the literature as *portfolio analysis* (Valdani, 1992), which by crossing growth rates, market shares and averages of production, exports and imports focuses on the dynamics of competitiveness in fruits and vegetables between the chosen countries/country groupings.

The results of past trends and salient characteristics of fruit and vegetables in DCs, as based on the aggregate data *supra*, raise a number of issues and questions-strengths, weaknesses and advantages for production and trade, possible impact of the general trend towards trade liberalization, implication of export competition, and the like. The statistical and economic analyses are therefore complemented with more detailed examinations of specific related issues, especially experiences of an individual country/product in the area under investigation. Concluding remarks and future prospects are reported in the final section *infra*.

Main Results and Related Issues

This section reports the main findings of the statistical and economic analyses. For reasons of space, only the main aggregates are considered, *i.e.* vegetables and fresh fruits and country groupings. Some comparisons with the aggreand country groupings. Some comparisons with the aggregate "World" (developing countries plus 8 areas of developed countries) are also presented for comparison. Comments on the trends and related issues are kept to the essential.

Vegetable Production

DCs account for 72% of worldwide vegetable production (1996). Over the time period 1986-95 production increased, on average, at a growth rate of 4.11%, a faster pace than the 2.11% world trend. Total production surged from 263 million metric tonnes (MMT) in 1986-90 to 327 MMT in 1991-95. All nine examined areas showed a positive trend, although the 6.6% growth rate for China was much higher than that for the other country groupings. By 1996 China accounted for more than 48% of the total vegetable production in DCs, a more than 10% rise from its 38% in 1986. South Asia was by 1996 the second leading producer at 19.4% of total vegetable output, followed by the Near East (8.48%) and East and South-East Asia (9.14%).

Of particular note here is the importance of China's top ranking in this cere is the importance of China's top ranking in this context and the impact it now has on the world horticultural sector as supplier, especially in the light of China's limits in storing and processing its domestic production. Among the other DCs, the developing vegetable production raises two priority issues, one related to income generation and the other to introduction of new types of vegetables.

Income Heneration

The evolution of vegetable production has an important economic significance in this area as output is relatively labour and capital intensive and vegetable growing by farmers is consequently dependent upon whether returns can at least cover production outlays. The role of vegetables in income generation is best illustrated through country examples.

In Vietnam much of the increased demand for fresh vegetables from the urban population, particularly for the more perishable species, is being met by peri-urban production, which has significantly increased since the change from a centrally planned economy with collective systems to a market-orientated one. Vegetables provide about $650 value added (returns to labour, land and management) per farm yearly for peri-urban vegetable farmers.

Guatemala provides a similar example. Vegetable production may be a promising option for meeting the two national goals of crop diversification and expansion of non-traditional agricultural exports because of rising foreign demand. Targeted to the smallholder sector, such production creates opportunities for labour-absorbing production techniques, although it also exposes farmers to greater risks related to market uncertainties. Production of export crops by small-holder farms in Guatemala appears to have had a positive effect on household income and food security, particularly for the smallest farms. Production of export vegetables increased on-farm employment, reducing the need to rely on uncertain off-farm employment in local and distant labour markets for additional income. New employment opportunities were also created for non-cooperative-member households. Disparities in income levels were also reduced (Von Braun, 1994).

Also in Malawi a wide range of exotic and indigenous vegetables are grown that play a critical role in the nutrition of the people by providing essential minerals and vitamins and generate some income for small-holders. This has prompted the Ministry of Agriculture more vigorously to pursue research, extension and personnely to pursue research, extension and personnel training in vegetable production (Mkamanga, 1990).

Another interesting example is in the North Indian state of Himachal Pradesh, where returns from vegetable growing are quite high and could be much higher if the package of recommended practices is adopted. Increases in vegetable acreage on small and marginal farms will not only provide gainful employment for under-utilized family labour but would also reduce income disparity among farms of differing sizes (Bathi, 1993).

New Types of Vegetables

The future of the vegetable industry in DCs will necessarily depend on improving people's standard of living as well as on the increasing, and more sophisticated, demand of consumers in large towns and cities for produce enhancement in terms of colour, taste, smell and shape. The new types of produce these countries, and China in particular, want to develop or to introduce from foreign countries are spicy vegetables, vegetable powder and convenience vegetables.

The demand for spicy vegetables depends on the fact that many people have become tired of mediocre delicacies but pungent spicy foods seem to be well-received all the time. Some food specialists believe that this type of produce can help dissolve fat and reduce its sediment in human tissues. Spicy foods are also said to be cancer human tissues. Spicy foods are also said to be cancer-resistant. Most vegetables can be reduced to powder after they are washed clean, the usual method being to pulverize, sterilize, dehydrate, dry and package (in plastic bags) them. The powder can be added to fermented dough before being processed into vegetable biscuits or noodles. It could also be made into instant vegetable soup, especially catering to the needs of those children or infants who dislike fresh vegetables. According to some food specialists, these vegetables show rosy prospects in sales and have become so popular in some big cities in the past two years that supply has fallen short of demand (USDA, 1998).

Packaged vegetables with a registered trade mark are growing quickly. The appearance of this type of produce in the marketplace indicates that many DCs will industrialize the mass production and processing of vegetables and establish a number of domestic and global name brands of their own. These products are thought to have a very bright future in the coming 21st century.

Production of Fresh Fruits

In 1996 DCs represented some 72% of World fresh fruit production. As with vegetables, China registered the fastest growth rate (7.61%) for fresh fruit production in all DCs, wherte (7.61%) for fresh fruit production in all DCs, where the average growth rate over the period 1986-95 was 3.25% (World: 1.46%). Production climbed from 253 MT in 1986-90 to 302 MT in 1991-95. Together with South America, China also garnered the largest market share with one-fifth (1996) of total fresh fruit production, the remainder being equally distributed in the other country groupings. China has a potential very similar to South America's, but with a much higher growth rate. The difference between the two areas consists in the fact that while South America is export-oriented towards the US and the EU, Chinese production is still earmarked for local consumption.

These trends raise at least two issues-the importance of the quality and stability of supply (China) and the role played by agricultural policy reforms (South America).

Quality and Stability of Supply

Fruit supply, in effect, should offer a minimum standard of quality and stability. Buyer surveys have clearly established the importance of quality, reliability of delivery and price as determining factors. Quality elements comprise grade of produce, defects, ingredients, mode of preparation of raw materials for processing, preparation and density of liquerials for processing, preparation and density of liquid medium, additives, drained weight, cleanliness, strength and fill of container and packaging material, and storage conditions to safeguard quality retention.

Although the fruit handling systems and marketing arrangements in DCs have evolved to take account of environmental factors and the lack of appropriate support systems, local marketing of fruit is still a high-risk enterprise. The fruit handling systems for mangoes and bananas clearly show the role of simple technological innovations, such as those designed for heat dissipation, reduction of physical injury or disease control, in reducing risks (Concepcion, 1994).

Infestation by fruit flies (*Tephritidae: Diptera*), common in the tropics, is a major constraint to the production and export of tropical fruits. Current control procedures include cover sprays of insecticides, spot sprays of protein baits, orchard sanitation and fruit wrapping, all of which seek to prevent direct damage to fruits or are aimed at population suppression, mostly in individual plantations. These field control techniques enable fruit production of sufficient quality to meet the needs of domestic consumption as well as that for export to markets where fruit flies are not quarantine pests. (Vijaysegaran, 1994). It is worth noting that evidence suggests that in the EU quality standards have suggests that in the EU quality standards have been more systematically applied against imported products rather than to what is domestically produced or marketed.

Diversification of Production and Agricultural Policy Reform

South America has benefited from a number of land reform measures, as well as from a general agricultural and fiscal policy encouraging the production and marketing of fruits for export. The fruit sector of several countries in this area has expanded. Chile, for instance, has developed this sector since the introduction of structural reforms in the 1970s aimed at market liberalization. Fruit exports have increased four-to five-fold, making Chile a prime example of boom in fruit production and exports. The development of the country's temperate fruit industry has been quite impressive-pears, grapes, apples, and the rapid introduction of new products such as kiwifruit in 1987. A similar evolution occurred in Argentina when in the 1980s the government began encouraging producers of cotton, sugar and tobacco to diversify into fruit and vegetables (OECD, 1996).

Exports of Vegetables

Exports of vegetables in DCs increased at an average annual rate of 4.91% over the period 1986-1995, going fro-1995, going from 3.6 MMT in 1986 to 5.8 MMT in 1995 when this area originated some 36% of World total vegetable exports. However, data presented show a strong variability according to the country grouping, *i.e.* in South America the export growth rate peaked at 18.71%, while South Asia registered a 1.85% decline. Central America is the main exporting area (37% of total exports in 1995), followed by the Near East Asia (18%) and China (13%). One issue that affects very differently South American vis a vis Asian countries relates storage and transport to trade.

Technical Progress in Storage and Transport vs. Trade Liberalization

The growth in exports indicates that in Central and South America exporters, processors and producers are quite capable of adopting technology and finding new export markets. Poor infrastructure, lack of food standards and grading, and importer concerns about food safety are constraints which will have to be addressed in Asian countries.

Although these problems can be considered as a limit to trade development, it is important to underline that trade liberalization has not developed at the same pace all over the world, and among the developing countries themselves the reasons are as much political as economic. A marked number of Asian DCs are not members of the Worldnumber of Asian DCs are not members of the World Trade Organization (WTO) and favour trade isolation. Export growth rates in most of the Asian country groupings are very low when compared with the corresponding level of production.

Vegetable Imports

On the import side the average growth rate of DCs has climbed to 5.15% (1986-95), going from 2 MMT in 1986 to 3.5 MMT in 1995 (World growth rate was 4.52%). In aggregate terms DCs represented in 1995 22% of World vegetable imports. The main importing areas were Near East Asia, with 37% of the DCs total import in 1995, and South Asia (33%).

The growth rate of vegetable imports is higher in South America with respect to Asian countries, where, however, import share is more consistent. The highly perishable nature of vegetables (as of all fruits) and the consequent need for immediate disposal calls for elaborate marketing arrangements in the absence of cold storage. South American countries have developed not only a good level of production but also an efficient storage and distribution system, and therefore can use raw products for the processing industry. In developing countries the vegetable processing industry has also important implications in relation try has also important implications in relation to employment.

Processing Industry and Employment

The processing of vegetables (as of fruits) helps to tackle the considerable potential loss arising from the perishable nature of the produce during times of seasonal glut in the context of low demand, poor infrastructures and weak transport. The reduction in tariff escalation helps to capitalize on off-season demand for high-value fruit in diverse developed country niche markets, enabling *horizontal* diversification away from traditional agricultural products.

At the same time, it allows more value added in the form of further processing and packaging rather than the mere export of raw products, thus opening up in the process possibilities of *vertical* diversification by generating both rural and urban employment.

Greater emphasis on a diversified and healthy diet has brought a substantial increase in the number and variety of fruit and vegetable products available to the consumer, with an ever increasing number of different types being processed and exported thanks to new heating (UHT, microwave, ohmic) and freezing (cryogenic) techniques combined with new packaging materials and technologies (aseptic, modified atmosphere packaging).

With the diminishing capacity of agriculture to provide employmentinishing capacity of agriculture to provide employment for ever-increasing surplus labour, the rural youth have been migrating to urban areas in search of a better life. The industrialization of rural areas is considered to be one of the most promising solutions to this problem. This is especially applicable to industries for the processing and preservation of fruits and vegetables (Khalil, 1993).

Export of Fresh Fruits

Fresh fruit exports by DCs (52% of World exports in 1995) grew at an average rate of 4.43% (World 4.32%), going from 14.2 MMT in 1986 to 18.1 MMT in 1995. Only North Africa showed a decline over this period (-2.61%). South America and Asian country groupings registered the most remarkable increases with a growth rate superior to 5%. In 1995 Central America covered more than one-third of fruit exports. With the share of South America (37%), these two country groupings originated some 70% of total exports by Developing Countries in 1995. Exports of temperate fruit and demand for tropical fruit products are two important issues affecting trade, with non-tariff barriers limiting potential imports from DCs.

Temperate Fruit Exports: Marketing Early or Late

Chile and Argentina are the largest exporters of temperate fruits (apple, pear, table grape and kiwi). To obtain favorable prices in export markets and to avoid tariff rates that otherwise could be very high, these countries take advantage of their seasonality, *i.e.* when their produce can be either early or late. Three types of methods are adopted to achieve this market advantage: cultivate early or late varieties (*e.g.* Chile harvests the early grape cvs. 'Perlette' and 'Flame' from early December essentially for the US market); adopt management techniques to bring harvest date forward *e.g.* plastic and hydrogen cynamide for grapes), adopt storage techniques to accelerate or slow the ripening process (*e.g.* controlled atmosphere, refrigeration and packaging film to slow and spraying ethylene to accelerate the ripening process).

During the last few years, sales by these countries have increased consistently. However, competition has also increased among them and with northern hemisphere countries. For instance, both Argentina and Chile are exporting the greater part of their pears to the EU, while the extension of the storage period for pears, currently nine months, has been an element of marked competition between producers in the northern and those in the southern hemisphere (OECD, 1996).

Demand for Tropical Fruits

Banana, Papaya, Guropical Fruits

Banana, papaya, guava and passion fruit deserve mention. The main banana suppliers are Costa Rica, Ecuador, Brazil, Honduras and India. Passion fruit producers are Ecuador, Brazil, Peru and Kenya. The major suppliers of mango are India, Peru and Ecuador, while guavas originate mainly from South Africa, Malaysia and India. The leading exporters of papaya are India, Peru and Brazil. For tropical fruits as a whole, the EU has been the fastest growing major import market during the decade 1986-96, with consumption more than doubling between 1984 and 1994. These fruits enjoy high income elasticities, *i.e.* a much higher consumption in proportion to the rise in income. A considerable amount of canned tropical fruit imported into the EU, however, is re-exported to other European markets, a substantial amount going to Central and Eastern Europe. Demand for tropical fruit products has, in recent years, risen on the international market, thus raising the stakes for certain developing country exporters.

Non-tariff Barriers

In general there are still non-tariff barriers that are limiting potential imports from DCs In this respect considerable importance is being attached to an international agreement on national sanitary standards and regulations. Maintenance of human and plant health are primary considerations in tropical fruit and vegetable-importing countries. The exporters have to satisfy consumer quality criteria as well as those of the relevant government agencies. The WTO's Sanitary and Phyto Sanitary (SPS) Agreement (separate from the Agreement on Agriculture) sets and encourages the adoption of international standards, while at the same time affirming the rights of countries to set their own health and safety standards if scientifically justifiable and not likely to result in an unreasonable barrier to trade.

Fresh Fruit Imports

Imports of fresh fruits in DCs grew at a yearly rate of 7.75%, doubling the amount traded (from 3.3 MMT in 1986 to 6.7 MMT in 1995). The share of DCs in the World aggregate went from 14% in 1985 to 16% in 1995. Remarkably, North Africa showed a 25.7% increase of imports, although the imports of this area account for only 1% (1995) of the total fresh fruit imports by DCs. Main importers of fresh fruit were the Nea6). Main importers of fresh fruit were the Near East Asia (34% of total 1995 imports) and East and South East Asia (29%) countries. It should also be noted that China is increasing its imports of fresh fruits (13% annual growth rate) and its share has grown from 3% in 1985 to 7% in 1995.

In Sub-Saharan Africa and South Asia the growth rate of imports is much lower because the capacity for self-production is higher. The relatively low growth rate in Asian countries is related to the high tariff duties on fruits.

The growth rates of imports by the various country groupings investigated are generally rather concentrated. This indicates that very differentiated production conditions are accompanied by similar trade situations, *i.e.* guarded opening to international trade. Until recently, most imports in South East Asia were channeled via Hong Kong, one of the area's few WTO member-countries.

The Underestimation of Chinese Trade Statistics

While China is a big importer of fresh fruit, official Chinese trade statistics vastly underestimate actual imports of fresh fruit as well as of many other commodities. Unrecorded shipments, rampant under-invoicing, misdeclaration of products, and other imponvoicing, misdeclaration of products, and other importation practices make Chinese trade statistics an inadequate indicator of real trade value and volume. Hong Kong's re-export statistics also reveal parts of China's import situation and usually are more accurate.

Hong Kong re-exports are imports that transit the territory on their way to other countries, including China. As an indication of Chinese customs undercounting of fresh fruit imports, Hong Kong re-export statistics show much higher apple, grape, and orange trade volumes and values than the official Chinese data. In China and in many Asian countries most imported fresh fruit enters through "unofficial channels", because of phytosanitary restrictions and high tariffs. Under-invoicing and misdeclaration of product are two of the most common unofficial importation practices. As an example of misdeclaration, traders will grapes or plums-prohibited by a phytosanitary barrier-as apples, which can be legally imported (USDA, 1998).

Per Capita Vegetable Consumption

Per capita apparent consumption of vegetables showed a 0.92% increase over the period 1986-95, despite the growth rate decline in

Africa and Central America. Apparent consumption went from 68.7 kg per capita in 1986 to 75.3 kg in 1995 on average. The highest consumption was registered in China The highest consumption was registered in China, where apparent per capita consumption jumped from 86 kg in 1986 to 146 kg in 1995 (a 5.38% growth rate). The lowest (and more stable) figure was registered by Sub-Saharan Africa (29 kg of vegetables consumed both in 1986 and in 1995).

New Consumer Trends in Vegetable Products

New trends in vegetable products for consumers in the DCs refers to people's standard of living. Generally speaking, fresh vegetables should lose market share to processed products. Many kinds of vegetables can be made into canned foods that cater to local tastes, cucumbers and peppers being but two examples. Easy to carry and convenient to serve, they can be stored for quite a long time and reduce losses incurred from seasonal supply surpluses of vegetables marketed at the same time of the year. At present the highest consumption has been registered by China, where three different kinds of vegetable should develop over the next few years-vegetable crisps, enriched vegetables and vegetable chips. Under vacuum-packaging at low temperature, radishes, melons and potatoes can be dehydrated and fried in edible oil into crisps. As they are processed in a low ambient temperature, which also reduced the loss of aromatic substances in their natural pigments, they can usually keep all their original flavors. They are regarded as a nall their original flavors. They are regarded as a new generation of food products upgraded to substitute the traditional fried and sponge foods (USDA, 1998).

On the other hand quite a number of scientists are doing research on so-called enriched vegetables, which contain many different nutrients. By means of conventional agricultural techniques, a certain number of trace elements can be dripped or sprayed onto the roots or leaves of vegetables to enhance them with the nutrients they lack. So far, Shanghai has developed a kind of vegetable called selenium cabbage.

Per Capita Consumption of Fresh Fruit

Per capita apparent consumption of fresh fruit has been relatively stable, with a growth rate of 0.38% (1986-95). Again, China has registered the highest growth rate (6.43%), while African and Near East Asian countries have shown a decrease.

It should be noted, however, that in China consumption is comparatively lower (54 kg per capita in 1995) with respect to the other areas, *e.g.* in Near East Asian countries apparent per capita consumption is three times higher. Apparent consumption went from 107.96 kg per capita in 1986 to 109.35 kg in 1995 on average.

Chinese Consumer Trends

It must be assumed that the growth in per capita consumption in the DCs is the result of China's marked increase. According to trade sources, Chinese consumers buy most of their fresh fruit at street retail shops and wet markets, *i.e.* approximately 95% of imported fresh fruit purchases are made at these locations. US and European fresh fruit has won brand-name recognition by China's consumers, especially Red Delicious apples, Sunkist® oranges and Red Globe table grapes.

Sunkist is one of the few overseas fresh fruit brands that consumers recognize. As an indication of its high level of local recognition, its packaging design has been imitated by a Chinese company with the intent to deceive local consumers. With the recent explosion in the number of supermarkets and hypermarkets, imported fresh fruit has begun to appear on retail shelves. While the amounts are still limited in comparison to those of the street stalls and wet markets, they could rise rapidly in the future (USDA, 1998).

Food Demand Projections

It should be noted in general that food demand in the DCs is projected to grow rapidly in the medium term due to a combination of factors-fast population growth, continuing rural migration to major urban centers and improved diets. Higher per capita income in these countred diets. Higher per capita income in these countries will provide impetus for a better and more varied diet, including to some extent fruits. Demand for fruit is generally very price elastic, thus final consumption is very much determined by the price, the quality and the disposable income of the consumer. (OECD, 1996, 1998).

Vegetable Exports and Imports

The total value of vegetable exports by DCs was US $3 billion in 1995 (28% of World aggregate), almost 50% of which originated in South America and 14% in China. The 1995 vegetable imports by DCs had a total value of US $1:4 billion (12% World aggregate), most of which originated from Asian countries (71%). The share of the East and South East Asian countries increased from 32% in 1985 to 42%

in 1995, while the Near East in Asia registered an opposite trend, declining from 42% to 29% of total vegetable exports.

For vegetables the 1995 standardized trade balance, an index showing the propensity of a given area towards importing (0 ÷ -1) or exporting (0 ÷ +1) a certain commodity (or commodity group), was close to 1 in North Africa (0.96), China (0 was close to 1 in North Africa (0.96), China (0.91) and Central America (0.89). Asian countries, although registering a negative standardized trade balance, over the same period of showed a tendency towards a decreasing import propensity, *i.e.* Near East Asia went from -0.36 in 1985 to -0.08 in 1995.

Fresh Fruit Exports and Imports

More than one-third of World fresh fruit exports originated in DCs, although over the long term their share decreased from 43% in 1995 to 37% in 1995. South America is the main exporting area (33% of total DC exports in 1995) followed by Central America and the Caribbean (27%).

In 1995 DCs accounted for some 17% of World fresh fruit imports, an increased share with respect to the beginning of the observation period (14% in 1985). In 1995 the main importers were Asian countries, which accounted for more than 60% of the total imports, and South America (17%).

Central America showed in 1995 the highest standardized trade index (0.82) in fresh fruits, followed by African countries (0.64 in North Africa and 0.50 in Sub-Saharan Africa) and South America and 0.50 in Sub-Saharan Africa) and South America (0.37). China registered instead a negative 1995 index (-0.15), its import propensity also declining with respect to the previous decade when the trade index was at -0.59 (1985).

Conclusion and Prospects

Over the past decade increased trade volumes of fruit and vegetables, together with significant changes in trade flows, have been indicative of a general increase in consumption and a greater demand for imports, with a corresponding general rise in production. Linked to these aggregate trends has been a variety of shifts in demand, production and trade flows both within and between Developing and Developed Countries.

A limited number of DC country groupings dominates production and trade. There is competition between them and their relative shares of world production and trade shifted from time to time. The limited number of competitive country groupings can be explained in part by the agro-ecological characteristics of the areas under investigation and in part by the sum of factors comprising the cost of production, processing, marketing, availability of appropriate technologies and infrastructure, product standards and technical regulations. The constraints to increasing horticultural exports from India, for instance, include poor Post-harvest management, inadequate cold storage and prpostharvest management, inadequate cold storage and processing facilities, high import duties on packing materials, underdeveloped market infrastructure, inadequate market information services, low levels of productivity and fragmented holdings (Subrahmanyam, 1994).

No less important are entry obstacles to export markets resulting from the specialized nature of the export-related infrastructure, including quality and health standards, as well as established consumer preferences for specific products in particular markets.

During the period under investigation, trade volumes were stimulated by a general trend towards liberalization as well as an extension of trading blocs. The general expansion in the demand and trade of fruit and vegetables has been accompanied by a greater internationalization of trade due to a greater number of importing countries as well as supplier countries engaged in international trade, which in turn has led to an evolution of trade flows. In many cases trade that was formerly local or bilaterally oriented is now much more international, which has intensified the trading contacts between Developing and Developed countries. The greater participation of the former in this trade evolution, both in the role of supplier and consumer, is thus considered to be a major development of the past decade in world fruit and vegetable production and trade. Strong growth of exports has occurrend trade. Strong growth of exports has occurred in the traditional exporting area of South and Central America (in particular Chile and Argentina but also Brazil and Mexico). This trend is expected to continue in the medium term. Significant growth in exports has occurred and is expected to continue in Asia (especially China and India).

During the period under review, the Asian region has been the area of stronger growth, which, given generally rising per capita disposable incomes and the vast potential of the Chinese market, should continue as a major growth market.

Apart from China, the wave of development is the result of the fast pace at which the so-called tiger economies of South East Asia have grown during the last decade. With a population of about 600 million, excluding China and India, this part of the hemisphere is ideal for the fruit and vegetable business from the quantitative point of view. Since the early eighties the countries of the region have passed through an unprecedented boom with 15-20% annual growth rate, stable currency and low wages. Unfortunately some countries have run up a great deal of currency debts and the boom has been followed by a heavy downturn. This does not mean that the area is no longer of interest for worldwide trade but certainly in the near future trade will suffer the lack of purchasing power in the most important countries of the area. This is a market which rematries of the area. This is a market which remains of great interest for mass-consumer goods like fruit and vegetables.

This trend is expected to accelerate both in the short and in the medium term as a result of the Uruguay Round Agreement, which will provide (a) increased market accession, (b) reduce subsidized export and aggregate levels of support, and clearer and more detailed rights and obligations for (c) food safety and phytosanitary regulations. Although at this stage it is too early to estimate the URA's impact on fruit and vegetable developments, and actually this will vary according to the country and to the product concerned, it is evident that these three elements will stimulate the trade environment.

The further extension of trading blocs such as the North American Free Trade Agreement (Nafta) linking Canada, Mexico and the United States, various association agreements in Central and South America: the Central American Common Market, the Andean Pact, the Caribbean Community and Mercosur, the Association of Southeast Nations (Asean: Indonesia, Philippines, Singapore, Thailand, Brunei), and the Lome Convention between the European Union and the ACP countries are likely to provide additional opportunities for increased trade.

Future prospects of the fruit and vegetable exports of DCs will also largely depend on the growth of import demand, mostly in the developed countries. Dort demand, mostly in the developed countries. Developed countries are expected to diversify their food consumption patterns, including consumption of horticultural products.

This heightened concern with health and nutrition and familiarity with more fruits and vegetables because of wider availability, increased travel and improved communications will lead to an increase in the ratio of imports to domestic products in total demand. Differences in per capita consumption of fruits and vegetables among developed countries with similar patterns and standards of living may also indicate a potential for increases in import-oriented consumption in countries where consumption is relatively low at present.

Look North to Canada as Promising Export Market

Trade in a foreign market doesn't always mean dealing with the Far East or Europe. A good place to start for U.S. exporters may be our northernmost neighbour—Canada. The country's proximity to the United States, and its common language and lifestyle, could make it a good place to gain valuable export experience and make sales. The U.S.-Canada Free Trade Agreement (FTA) awakened many small- and medium-sized U.S. food companies to the fact that an important export market was right next door. Canada is the second largest market for U.S. agricultural exports and U.S. exporters are the dominant suppliers of agricultural exports to that country.

Of the $4.8 billion in U.S. agricultural exports to Canada in 1992,(*) 75 percent were in the high-value consumer foods category. In addition to fruits and vegetables and their products, the United States is the major source of other high-value consumer foods, such as beef, poultry, bakery products and pet food. Corn, soybean meal, vegetable oils, feeds and fodders and sweeteners are traditionally the major bulk and intermediate products exported to Canada.

Changing lifestyles in Canada are increasing the demand for U.S. agricultural products. With more women in the work force, there has been a reduction in the number of meals consumed at home. This has led to increased imports of U.S. consumer-ready products. Also, consumer awareness of health, nutrition and calorie-reduced foods, along with interest in international foods, has accelerated demand for U.S. packaged foods.

Fruit and Vegetables

Chile: Leader in Southern Hemisphere Fruit and Vegetable Exports. In the 1950s, the fruit and vegetable industry began a seedy export-oriented growth trend. Technological developments were the determining factor in launching this important production industry into the international arena.

Chilean fruit and vegetable exports date back to the early 20th century; however, the industry's steady growth did not begin until the early 1950s. The main factors that promoted this development were technological advantages that allowed shipping the products in vessels with cold storage chambers. The fruit and vegetable industry began in the late 1970s and early 1980s and it was due mainly to two factors:

- Chile 's natural conditions for producing fruit and vegetables that grow in mild climates.
- Changes in Chile 's economic policies, based on a strategy involving participation and opening up to foreign trade.

Advantages for Fruit and Vegetables

Chile enjoys a series of advantages for producing fruit and vegetables:

- Favorable soil and weather conditions
- Ideal phytosanitary conditions thanks to its geographic isolation originating in natural barriers (the Pacific Ocean, the Andes Mountains, the Atacama Desert and Antarctica).
- Counter seasonal production in contrast with the main consumer markets in the Northern Hemisphere.
- High standards among entrepreneurs and a qualified labour force, which translates into the availability of a skilled labour force.
- Low and rational use of agrochemicals. Economic policies that favour free enterprise. Diversified and staggered production.
- Constant introduction of new varieties, production technologies and preservation of fruit and vegetables, added to adequate shipment methods (sea, air or land) that ensure an uninterrupted year-round supply.

Evolution of Exports

Shows how fruit and vegetable exports have evolved, from US$168 million in 1980 to US$1679 million in 2003, *i.e.* an almost 9-fold growth in a period of 23 years.

Table: *Evolution of Fruits and Vegetables Exports*

Year	*US $ millions FOB*	*Thousand of tons*
1980	168	261.0
1985	356	533.0
1990	716	1,034.6

Contd...

Year	*US $ millions FOB*	*Thousand of tons*
1995	1,162	1,310.1
2000	1,386	1,506.4
2001	1,359	1,684.7
2002	1,534	1,830.5
2003	1,679	1,980.2

Source: *Chilean Central Bank*

In the mid 1960s, Chile 's share of all exports of grapes, kiwi fruit, apples and pears from the Southern Hemisphere amounted to only 3.5 percent. At present, Chile is the market leader in terms of volume, and it holds a 48 percent market share with regard to these species.

Chile is the main worldwide exporter of grapes and the second largest exporter of apples and pears.

In the Southern Hemisphere, Chile is the largest exporter of apples, peaches, berries and avocados, ranking second in pears and kiwi fruit.

Notably, 24 percent of the world's grapes and 9.1 percent of the world's peaches are produced in Chile. They are harvested when these fruits are out of season in the Northern Hemisphere, thus making Chile the main exporter for these products in the opposite season.

In the case of vegetables, specifically, average exports surpass US$31 million and the total volume is around 68,000 tons.

The main vegetables exported by Chile are onions, garlic, radicchios and asparagus. A comparison of fruit and vegetable exports in 2003 with those made in 2002 shows that export amounts increased by 9.5 percent while export volumes grew by 8.2.

There are major changes in terms of the varieties of fruit grown and exported. For instance, Crimson Seedless is a new variety of table grape recently introduced that is expected to continue growing. Traditional groups such as Delicious apples have gradually been replaced by two-coloured varieties for which demand is currently rising, such as Gala, Royal Gala, Scarlett, Fuji and Braeburn.

Table: *Fresh Fruits and Vegetables Exports*

	Volume (Thousand of tons)				Value (MM US $ FOB)			
Species	2000	2001	2002	2003	2000	2001	2002	2003*
Grapes	596.2	545.3	654.9	730.0	662.5	580.4	645.6	710.0
Apples	387.7	542.7	548.2	585.0	202.2	237.2	275.1	305.0
Pears	124.8	128.8	119.4	122.5	64.4	60.1	63.0	68.0
Kiwi fruit	102.5	110.4	117.5	107.0	68.3	65.9	86.7	85.0
Plums	52.7	81.6	75.9	80.8	64.8	70.7	68.9	70.0
Nectarines	43.8	52.0	51.1	58.5	44.8	44.8	44.5	46.0
Peaches	30.8	37.7	37.6	44.4	25.9	35.3	32.9	35.0
Cherries	6.1	7.5	12.8	11.0	22.6	29.6	37.5	35.0
Avocados	52.0	52.5	78.1	80.0	73.8	79.6	97.4	110.0
Raspberries	3.6	4.1	4.3	4.5	21.1	21.3	23.2	25.0
Blueberries	4.0	4.4	6.4	6.4	12.7	13.9	18.1	24.9
Lemons	18.0	21.0	25.9	30.0	14.1	20.0	31.1	35.0
Onions	29.3	34.7	39.4	51.0	7.9	10.4	10.3	15.0
Garlic	10.3	8.9	10.7	11.0	9.1	8.5	11.7	10.1
Radiahios	2.3	2.3	2.5	2. 8	4.2	3.7	2.6	2.6
Asparagus	3.6	2.9	2.1	1.7	5.5	3.9	2.9	2.0
Other fruit	26.5	33.5	41.4	52.1	74.6	66.1	80.6	98.8
Other vegetables	12.2	16.4	2.3	1.5	7.1	7.9	1.7	1.3
TOTAL	1,506.4	1,684.7	1,830.5	1,980.2	1,385.6	1,359.3	1,533.8	1,678.7

Source: *Chilean Central Bank/*2003 Forecast*

Other Prominent Species

Species with considerable expansion rates in terms of export volumes include avocados, blueberries and lemons. The underlying reason is an increased surface area farmed in the last few years.

Main Species and Destination Market

Volume (Tons) - 2002/2003 Season						
Species	North America	Latin America	Europe	Far East	Middle East	Total
Grapes	416,217	80,583	129,473	65,650	6,334	698,257
Apples	95,348	202,808	203,945	30,429	65,349	597,879
Pears	19,604	32,113	56,088	716	4,698	113,219
Kiwi fruit	20,508	11,194	72,385	8,180	1,517	113,856
Plums	31,293	15,006	20,688	9,757	1,287	78,030
Nectarines	37,746	9,147	8,379	1,644	86	57,003
Peaches	29,615	15,782	1,107	175	0	46,679
Cherries	8,352	1,410	1,380	2,743	2 8	13,284
Avocados	89,297	123	2,687	8	0	92,115
Blueberries	4,877	7	372	216	0	5,474
Lemons	15,133	0	219	14,026	0	29,378
Other Fruit	90,707	8,899	18,094	9,137	778	54,070
Total 02/03	784,597	377,072	514,817	142,681	80,077	1,899,244
Total 01/02	671,426	440,881	437,592	128,145	74,670	1,752,714
Percent Variation	16.9%	-14.5%	17.6%	11.3%	7.2%	8.4%

Source: *Chilean Exporters Association (ASOEX)*

At present, Chile makes direct export of more than 75 fruit and vegetable species to 70 markets. The main species exported are table grapes, apples, pears, kiwi fruit, pitted fruit and avocados. As shows, the most important species is table grapes since it accounts for almost 36.9 percent of the total volume of exports, followed by apples with 29.5 percent of the total volume.

Markets for Fruit and Vegetables Export

In the 2002/2003 season, the main destination was North America (United States and Canada) with 41.3 percent of shipments sent there, followed by Europe and Latin America to which 27.1 percent and 19.9 percent, respectively, of the shipments were sent. As shows, the market share in other regions was Far East 7.5 percent and Middle East 4.2 percent.

Benefits of Free Trade Agreements for the Industry

The advantages of the Free Trade Agreement between Chile and the United States include the strengthening of the Generalized System of Preferences (GSP), a reduction in the high duties applied to avocados and asparagus, the establishment of a dispute settlement system, and a special committee on phytosanitary matters. In the Agreement between Chile and the European Union (FU), 85 percent of fresh fruit now enjoy tariff free status. This places Chile in a favorable position versus its competitors. Fruit and vegetables with high potential such as apples, pears, plums, avocados, raspberries, onions, asparagus and bell peppers can already enter the EU markets under full customs duty exemption.

Table grapes enjoy a benefit consisting of free quotas during two very important export periods: from January I to Icily 14 and from November 1 to December 31 of each year. Garlic was granted a 500-ton quota, with an annual increase of 25 tons.

Characteristics

Of all the different sub-sectors within the food industry, fresh fruit and vegetables have a chain of supply which displays unique characteristics;

- Fruit and vegetables are labour-intensive horticultural crops, mainly grown by small family enterprises. However, also banana companies and large USA fruit and vegetables enterprises are active in the market.
- In order to increase market power many of these smaller growers have organised themselves into sales organisations.

- Fruit and vegetables are traded as a fresh perishable products, makinlves into sales organisations.
- Fruit and vegetables are traded as a fresh perishable products, making the time factor and packaging critical.
- Nearly every single country in the world is involved in the trade resulting in a very large number of exporting and importing countries and a wide range of companies.
- The range of fruit and vegetable products is becoming increasingly wide.
- The range and origin of supply may shift from season to season. This unique situation of fruit in combination with vegetables, created in the retail product range makes it difficult for suppliers and retailers to establish agreements on long-term year-round supply and prices.
- A large number of retail outlets and formats exist, varying from large retail chains to small restaurants.
- Added value in the fruit chain is limited. This restricts the opportunity to create brands. The lack of clearly distinct consumer brands makes it easy for the buyer to shift from one supplier-trader-to another, and suppliers are therefore vulnerable. Buyers' loyalty to traders is based on service rather than on the branded products they sell.

Number and Size of Holding

The structure of the fruit and vegetable exploitation is not homogeneous. The complexity reflects the diversity of factors: evolution of consumption, production techniques, socio-economic interest, general politic climate, production policy, international competition and international trade agreements competition and international trade agreements.

Eastern European countries and many developing countries, still have a predominant traditional family-linked production unit. Traditionally fruit and vegetables were exclusively cultivated close to the cities, particularly for local consumption. Later transport over larger distances permitted specialisation and the de-localisation of production. Historical reasons explain the actual co-existence of different production systems:

- domestic production, particularly destined for auto-consumption, important in developing countries and in the

Mediterranean area. Industrialised countries have garden plots and private orchards in areas where the climate is favourable for natural production.

- agricultural farms not specialised in fruit and vegetables but having a garden plot or orchard in order to obtain an additional income.
- professional enterprises, specialised in fruit and/or vegetables. A distinction is to be made between family holdings who use hired labour force for seasonal workers and professional non-family enterprises of larger size utilising one of more permanent employed persons and a large number seasonal manpower.

According to the results of structural analysis in EU in 1989-1990 approximately 1.8 million of farms produce fruit and vegetables. Surprising is the small average dimension: 1.3 ha.

However, the analysis of professional exploit.3 ha. However, the analysis of professional exploitations shows that the average size is considerably larger. There are about 100.000 exploitation's specialised in commercial production of vegetables and 350.000 specialised in the commercial production of fruit. The average size is 4,2 ha for specialised exploitation's in vegetables and 7,9 ha for specialised exploitation's in fruit.

The specialised holdings achieve about 50% of the commercial production of the EU for vegetables and over 80% for fruit. The technological progress particularly in biotechnology and information did increase the operating scale and profitability of the exploitations. Particularly the integration of technological renewal in horticulture is important. The entrepreneur has to adapt his production methods and management in order to increase cost efficiency. The USA census 1992 reports 99.514 farms producing fruit and nuts, 29.605 vegetable (incl. melon) farms and 39.712 other specialised horticultural units. The average size of the USA farm is 21 ha for fruit and nuts and even 50 ha for vegetables and melons.

As compared to the EU, the USA 'family farm' dimension is significantly different. More-over, typical for both USA and EU, about 25 percent of the farms produce about 75 % of the production.

The requirement of capital per exploitation and per full-time employed person decreases per unit of the production. Thed person decreases per unit of the production. This evolution of production factors leads to the suggestion that the need for capital will increase and the need for labour will decrease. This is logically followed by a decrease of the number of exploitations. It is clear that the need for knowledge, comprehension and management capacities at the farm exploitations are still increasing.

Income Generation

Most fresh fruit and vegetables are produced by small and medium sized family farms. Once the seeds or trees have been planted, costs can be kept very low, particularly when labour is to a large extent supplied by the family itself. This makes the supply of fruit and vegetables comparatively less susceptible to price fluctuations; farmers will continue to supply the market, even when prices are only slightly higher than variable costs for several years. In the longer term, however, higher prices are needed to allow farmers to replant especially as planting new varieties is becoming increasingly important supply may vary greatly from season to season depending on the weather. According to economic theories a short supply would result in higher prices, which in turn counterbalances low volumes and therefore levels off the income effect.

Fruit, however, from one destination or one kind may easily be substituted by fruit from other sources or kinds. Consumers may easily shift from one fruit to ainds. Consumers may easily shift from one fruit to another when prices for a certain fruit would be considered too high. Therefore producers and traders will rarely benefit from very high prices in years of shortages, and they will continue to experience the disadvantages of very low prices in years of abundance. This market situation with either low or very low prices results in structurally low margins for all parties. As a result of the price and marginal situation in the market, producers and exporters are constantly searching for possibilities to earn premium prices, and the role played by technology in these and other developments in the fruit chain is becoming increasingly important.

Supply

Worldwide, the 1996 production (FAO) reached 508 million tons for vegetables and 469 million tons for fruit. The productions dynamics is different according to continents, countries and types. Production is developed countries showed a strong expansion in the period 1950-60, but is stabilising. The yearly production growth in developing countries is more important than the demografic growth. This expansion matches the growing demand in Asia, Latic America and the Northern countries (out of season, tropical etc.)

Vegetable production increased by 56% over the last 15 years. Asia dominates with 60% of the world production. The EU comes next with 11%. The most importanroduction. The EU comes next with 11%. The most important producing countries are China, India, EU and USA. Over 100 types of vegetables are produced and consumed in the world. Only the types having good taste qualities and allowing for technical progress with respect to production and marketing are developed on a world scale. The breeding of vegetables, for example, is especially important for varieties cultivated in greenhouses or destined for further processing. This may result in standardisation of production.

Table: *Production Volume, 1000 MT*

	AV 93-95	GR % 87-94
Citrus	87.933	3,48
Bananas	82.217	2,74
Tomatoes	80.664	2,94
Grapes	55.299	-1,67
Apples	48.920	2,68
TOTAL FRUIT(*)	448.588	1,57
TOTAL VEGETAB/CENTER>		
TOTAL VEGETABLES	489.242	3,21

**Excluding nuts*

The development of the fruit production shows a similar trend to that of vegetables, having recorded substantial growth from just over 300 million tons in 1980 to in excess of 400 million tons in 1995, an increase of more than 30% over the last 15 years.

Growth rates varie largely. Grape production decreased by 18% during this 15-year period, whereas production of citrus, bananas and apples increased by 50%, 48% and 44% respectively.

The fall in grape production is quite an exception since most types of fruit have shown considerable growth. The largest growth rates were achieved by relatively smaller fruits, with the kiwi fruit having established itself as a major fruit following a rise in production of more than 3000% in the past 15 years. The more luxury fruit types have also experienced large growth: sour cherries and blueberries, for examples have shown growth rates of 102% and 98%. Tropical fruits have performed very well too, papayas growing by 81% and mangoes by 36%. Apart from mangoes though, all these fruits are still relatively minor components of the total fruit package. The above average production growth of the luxury and tropical fruits is a response to more sophisticated consumer demand in the developed ecosophisticated consumer demand in the developed economies that requires a wider variety of fruit throughout the year.

***Table:** Production Value US $ 1.000.000*

	AV 93-95	GR % 87-94
Citrus	22.976	3.7
Bananas	25.148	4.64
Tomatoes	31.902	5.13
Grapes	31.945	1.55
Apples	14.377	4.54
TOTAL FRUIT	134.599	1.77
TOTAL VEGETABLES	159.237	5.25

It is not only demand that has given momentum to fruit production, but policy has also stimulated fruit production, either directly or indirectly. Changes in the agricultural policy of many countries have led to farmers diversifying production and many have opted for cash crops sifying production and many have opted for cash crops such as fruit and vegetables. In North America, for example, the support for cereals and oilseeds was reduced in the mid 1980s, and this resulted in a subsequent increase in acreage under fruit trees in certain areas as well as diversification towards fruits of a higher added value such as berries. Chile is another example where structural reforms introduced in 1974 gave farmers greater security on their farmland property rights. Prior to 1974, annual crops tended to be grown because of the farmers' unwillingness to invest. But allowing farmers to own the land provided the incentive for them to invest in perennial crops and, as a consequence, fruit production boomed.

Supply Policy USA and EU

EU Market Management

More than in the other agricultural sectors the EU production of fruit and vegetables is guided by the market forces and influenced by tariff barriers. Therefore, for most products the surplus of production is minor.

There are however exceptions to this general rule. The EU withdrawals (below) are important and did become routine for a number of varieties of apples and peaches. Also in the case of citrus the withdrawals are relatively small but a growing part of the production is not suited for the fresh markets and is diverted (without support) to processed products.

The interdependence of producers of fruit and vegetable. The interdependence of producers of fruit and vegetables on the market influences the future evolution: if the internal market becomes oversupplied, a price drop will follow and lead to a progressive adjustment.

An external factor of increasing concern is the reform of the Common Agricultural Policy. A number of cereal producers are attracted to diversify their production to fruit and vegetables. They take advantage of the larger size of their enterprices and are supported by the EU for their traditional production. A similar situation is reported for vineyards, taken out of production by EU subsidies ('grubbing') and converted to production of apples.

The European withdrawal system was designed to cope with the cyclical surpluses unavoidable owing to the seasonal variations in yield due to weather conditions. Up until 1968 the withdrawal was entirely financed by the producers organisations. Modelled on the Dutch experience, the EU experimented with various forms of market management including :

- leaving the fruit on the tree and the vegetables in the field;
- setting increasingly demanding marketing requirements (*e.g.* minimum sizes);
- destroying part of the crop.

In practice the intervention by the EU was limited to setting the basic price and buying in price for the following produce only: eating apples, apricots, dessert pears, peaches, nectarines, table grapes, citrus, toert pears, peaches, nectarines, table grapes, citrus, tomatoes, cauliflower, aubergines, courgettes and cucumbers.

A preventive withdrawal system has been in operation since 1979 for apples and pears as these can be stored. This mechanism was introduced as it became apparent that it was better to prevent a market collapse than to react once a crisis had occurred. In 1987 the EU begun to implement stabilising measures because of the fact that the cyclical events turned out to be a structural characteristic. Indeed, almost the entire crop of white peaches and red oranges ended up in withdrawal from the market. The withdrawal system was gradually loosing the initial aim of curing cyclical aspects. Consequently, the budget was rising to approximately, one million ECU on a yearly basis.

Moreover the public opinion became very dissatisfied with the size of the market withdrawals and particularly with the destruction of lots of products. The 1996 EU-reform of the fruit and vegetable market organisation is aimed at bringing a better equilibrium between supply and demand in the European Union. Budget saving are not the first goal. But the actual amount (2 billion ECU) would better be spent in order to satisfy producer and consumer.

USA Market Management

Most crops are subject to federal (USDA) or state quality standard. Both shipper and buyer are familiar with the USDA requirements shipper and buyer are familiar with the USDA requirements to be met for certain grades or sizes. If, on receipt of the product the buyer is satisfied that it is the requisite quality level, the product is accepted without any government inspection or involvement. The non-mandatory quality standards are simply there to facilitate trade, keeping direct costs down. One of the most important sources of institutional support for the US fruit and vegetable industry are mandated marketing programmes. Growers are free to choose from a menu of legal organisational options designing the programme that best meets their needs:

- Quality control improving the product sales and volume management provisions such as permitting only a certain portion of the crop to move into specified outlets (*e.g.* producer allotments);
- Market facilitating provisions such as production research, market research and development, market information;
- Market promotion and advertising.

All of the firms producing, and in some cases handling the commodity in a region encompassed by a mandated programme are required to pay an assessment levied on each unit sold. The funds are then pooled and administered according to the purpose agreed upon by the commodity group. The interest of the growers and/or handlers are represented by their elected grower and/or handler boards of directors.

Growers and/or handlers continue to sell their produc Growers and/or handlers continue to sell their products independently as usual, merely contributing assessments on each unit sold. Promotional programmes are always conducted on a generic basis-promoting the product rather than any particular brand -.

Fruit and vegetable orders are active for: Florida citrus, limes and avocados, Idaho-Oregon onions and South Texas onions, all approved prior to 1991. It has been many years since new marketing orders with quantity control have been approved and it is viewed as highly unlikely that any will come into being in the future.

State marketing orders are required to fund production research through the corresponding state (Land Grant) universities. Research and promotion funds for fruit and vegetables collected under federal marketing order and check-off programmes are estimated to have totalled $ 45 million annually. The total farm value of the US fruit and vegetable industry was $ 23.000 million in 1994. Hence in relative terms, the industry investment in research and promotion through federally authorised programmes amounted only about 0.2 percent of the total cash receipts.

Importance of Producer Organisations

A new EU market management system was introduced in 1996. The institutional withdrawal prices are reduced gradually by 40 percent (30 percent for apples and pears) over a period of six years recent for apples and pears) over a period of six years. The withdrawal prices will be fixed for the entire EU as from 2002 on.

The EU 1997-reform is aimed at promoting the producer organisation

Key elements of the reform are :

- new responsibilities of the producer organisations;
- operational programmes;
- reform of the withdrawal system;
- interprofessional organisations;
- respect of environment.

The main objectives of producer organisations are:

- to market all the products produced by its members and to make sure that the production is planned and adapted to the consumes demand in the market, especially concerning quantity and quality; producer organisations receive EU funding on a 50/50 basis in order to realise operational programmes:

- to improve the quality of the products, to increase the commercial value of the products, to encourage the consumption, to develop the biological production, to encourage integrated production and to limit the withdrawal from the market;
- to take measures to encourage the producers to apply environment friendly production techniques, for the production as well as for the management of inputs;
- financial planning in order to apply the quality standards, the phytosanitary regulations and the maximum residue levels.

The creation of agricultural cooperatives in the USA is supported by law. Despite the appatural cooperatives in the USA is supported by law. Despite the apparent appeal of cooperatives, they are no longer a major part of the institutional setting in the fresh fruit and vegetable industry. Marketing cooperatives have been more important to the processing than the fresh market crop and particularly to the citrus sectors. In general the declining appeal of cooperatives is probably partly related to the increase in average grower size, as larger growers are able to market their crops independently.

Supply of Some Important Crops

Citrus

World citrus production totalled 87,9 million tons in 1994, 23% of all fruit produced. It can be seen that the gradual increase in production occurred mainly after 1985. For grapefruits, most of the growth occurred during the nineties, a rise that can be explained by the development of new, less bitter varieties that are better adapted to the taste of consumers.

Differences exist between oranges grown for processing or for direct consumption. Fruit sold on the local fresh fruit market often yields a better price than fruit sold to the processors, fresh fruit demand needs to be satisfied before oranges are supplied to the processing industry.

Encouraged by good producer prices, many orange trees were planted in Brazil between 1988 and 1993 but lower prices in recent years have discouraged the planting of new grapes and may curtail priscouraged the planting of new grapes and may curtail production growth in the future. In the USA citrus production is moving south within the state of Florida to reduce the risk of frost.

In Mexico and Italy most of production is consumed domestically. In Italy and the USA 8% of citrus production is used for export purposes, whereas Spain is the only country which mainly produces for the export market with more than half of production exported.

Table: *Citrus Production, 1000 MT*

	Production AV 93-95	GR % 87-94	
Africa	7.423	1,80	
Latin America	30.780	3,66	
Asia	22.137	7,33	
Industrialised Countries	-	-	
EU	9.161	1,25	
NA	13.828	3,55	
TOTAL	87.933	3,48 </T3< CENTER>	3,48

Grapes

The fall in grapes production over the past 15 years is evident. With more than 70% of production used in wine making, the largest part of the production fall is attributable to the fall in wine production. This is clearly illustrated by the fact that grape production declined especially in Italy, Spain and France, the world's major and traditional wine producing country. Europe is responsible for 72% of world wine production, 52% of grape production and 58% of vine area. In the USA, Turkey and Southern Hemisphere countries a relatively high share of grape production is used for raisin production and fresh use.

In contrast with total grape production, the production of table grapes has remained rather stable during the nineties and is estimated at ten million tons. Production of table grapes is concentrated in four countries with Turkey, Italy, Chile and the USA (California) having a global market share of more than 60%.

Table: *Grape Production, 1000 MT*

	Production AV 93-95	GR % 87-94	
Africa	1.290	0,14	
Latin America	5.414Latin America	5.414	-0,33
Asia	9.802	2,96	
Industrialised Countries	31.877	-2,06	
EU	23.506	-2,97	
NA	5.436	0,96	
Economies in Transition	6.916	-4,73	
TOTAL	55.299	-1,67	

Bananas and Plantains

World banana production increased from 43 million tons in 1978 to 54 million tons in 1994, an average increase of 3,12% per year. Ranking second and fifth among the world's major fruits, bananas and plantains are important crops. The plantain or cooking banana is a variety of banana which is part of the daily diet in tropical areas. With less than 1% of total production exported, the product plays an insignificant role on the world market.

Because of the banana's dual character, being produced both as a staple food and for export racter, being produced both as a staple food and for export purposes the production structure is very diverse, often very small scale and not on a commercial basis because a banana tree can be grown in every suitable backyard. Banana production for commercial purposes varies from smallholder production with farms often smaller than five hectares to production on plantations of several thousand hectares. The scale of banana production is to a large extent determined by the natural conditions.

Table: *Banana Production, 1000 MT*

	Production AV 93/95	GR % 87-94
Africa	6.581	1,77
Latin America	22.870	3,49
Asia	23.119	3,39
Oceanic	687	3,03
Economies in Transition	827	-1,86
TOTAL	54.083	3,12

In Central America, where most of the dollar bananas originate, production is fairly export oriented. Independent producers exist, but a considerable part of production is produced on plantations owned or controlled by one of the large multinational banana companies, the best known of which are United Fruit, Dole and Del Monte. The large-scale plantations, sometimes as large as 5.000 hectares, are characterised by efficient organisation, substantial know-how, fast implementation of innovations, high yields, high consistent quality and a relatively high capital input. Independent producers tend to dominate production in EU and ACP countries, varying from small scale production to large plantations. Most Caribbean producers have plantations of 5 to 20 hectares, though in some countries larger plantations exist, up to sizes of 2.000 hectares.

Exports from these countries are mainly organised through independent European importers (Fyffes, Pomona) Independent European importers (Fyffes, Pomona).

Apples

The average annual world growth is estimated at 2,68 percent, particularly because of the production increase in Main China. The world apple production for the year 2000 is estimated (Washington State University) at 62 million ton. Red Delicious is world wide the most important variety, followed by Golden Delicious. Fuji has a remarkable recent increase: 80 % of the total acreage of this variety, namely ncrease: 80 % of the total acreage of this variety, namely 370.000 ha, is planted in China. China had only a small production in 1987 and is now the world largest producer.

It is expected that the acreage of red varieties will remain stable in the year 2000 and so does the acreage of Golden Delicious. The acreage of bicolour varieties will likely increase. Red Delicious remains the number one variety for the USA. The share of Granny Smith, Empire, Fuji and Gala is increasing. Plantings of Pink Lady are started. Production in EU the still increasing but there has been a shift from green and yellow varieties to red and bicolored varieties. Jonagold, Gala, Breaburn and Elstar are trend-setters. Standard Golden Delicious, Red Delicious, Granny Smith, Cox's and James Grieve are decreasing.

It seems that some regions are looking for typical varieties such as Jonagold in Belgium, Elstar in Holland, Golden Delicious in Limousin and Pink Lady in South France. There has been a switch to fruit growing by cereal farmers and wine grape growers as a result of lower guarantee prices for cereals and wine. Large scale intervention has occurred in the EU in bumper crop years (Sec 2.1.1.). Germany and Switzerland are looking for new resistant varieties particularly on farms with direct selling to consumers.

Eastern Europe is looking for varieties, suited for the Western European markets, such as Jonagold, Elstar and Pinova. But thearkets, such as Jonagold, Elstar and Pinova. But the majority of the orchards are out of date. The renewal is relative limited because of lack of capital, uncertainty on land-ownership, *e.g.* in the former Eastern Germany. In the Southern Hemisphere, Argentina used to be the most important producer. The SH has plenty of possibilities but the capital is lacking. Today Chile is the most import apple producer in SH with a relative decrease of green varieties and an increase in

Gala, Braeburn, Fuji and Scarlett. Some parts of Brazil are climatologically well suited for Gala.

Granny Smith is the main variety for South Africa, occupying 50 % of the plantations: the share of Gala is increasing. South Africa has the possibility to increase the production volume by 25%.

The production in New Zealand is stabilising after many years of increased renewal of older plantations has taken place. The Apple and Pear marketing Board has an important role in variety selection and quality policy; Pacific Rose is considered promising.

Table: *Apple Production, 1000 MT*

	Production AV 93-95	GR % 87-94		
Africa	851	15,97		
Latin America	3.213<>	Latin America	3.213	3,52
Asia	18.815	9,69		
Industrialised Countries	17.833	0,47		
EU	9.563	-0,62		
NA	5.498	+2,42		
Economies in Transition	8.209	-3,74		
TOTAL	48.920	2,68		

Tomatoes

Over 80 million tons of tomatoes, 19% of the fresh vegetable production in the world, are produced every year. The world production of tomatoes only was 35 million tons in 1970.

This increase resulted in part from the increase of area but for a more important part from technological innovations and the development of productions in glasshouses. Many countries are producers of year-round tomatoes and not only in the summer-months, using new to varieties adapted artificial heating and high technology.

Tomatoes are produced and sold in all zones of the world. Theatoes are produced and sold in all zones of the world. The EU is the most important producer with 13 million tons of tomatoes or 18% of the world production.

The Mediterranean areas of the EU (Italy, Spain, Greece, Portugal, South-France are the most important producing areas. However an important part of the crop is produced by northern European countries (Holland, Belgium). The USA is the next important producing country. Tomatoes are the USA third important product after citrus and grapes. California and Florida are the two important producing regions.

The Turkish production however has an exceptional growth rate going from 1,2 million tons in 1961 to 6 million tons in 1994. This is particularly due to the development of irrigation. The total irrigated area in Turkey represents 20% and an additional increase is to be expected.

Approximately 30% of the world production is destined to further transformation.

Table: *Tomato Production, 1000 MT*

	Production volume	GR % 87-94		
Africa	8.732	2,53		
Latin America	7.584	2,47		
Asia	29.547	Asia	29.547	6,69
Industrialised Countries	26.914	2,7		
EU	12.792	1,43		
NA	12.021	4,46		
Economies in Transition	7.882	-4,62		
TOTAL	80.664	2,95		

Trade

Overview

Within the chain of supply, there are several trade steps between the fruit grower and the end consumer. A distinction is made between:

- Collection wholesaling (collecting the fruit from the grower).
- International trade
- Distribution wholesaling (distributing the fruit to the retail outlet).

A number of traders combine two or even three of these functions. Almost all vegetables and most of fruit in the world are consumed as a fresh product in the domestic market where domestic traders are active.

Indeed, some of the largest fruit and vegetable growing nations produce almost exclusively for the domestic market. However, over 30 % of frumost exclusively for the domestic market. However, over 30 % of fruit is processed mostly into juice and wine, and these are often traded internationally as well (juice is traded predominantly in a pre-processed form as concentrate).

While fresh fruit is consumed all over the world no type of fruit can be grown everywhere. Since consumers demand a wide choice, trade must take place between producing and consuming regions, making the trade function particularly important. Nearly 9 % of fruit production is traded internationally as a fresh product. The proportion of fresh fruit which is traded increased from 6,9 % in 1980 to 8,9 in 1995. While this is low compared with other agricultural commodities such as wheat, it is high compared to vegetables which are predominantly grown in consuming areas.

Because fruit can be consumed both fresh and in processed forms, the volume available for international trade depends on a combination of factors including annual production, levels of processing and domestic consumption. Competition between these market segments, however, is relatively low as each segment consumes produce of a different quality.

Table: *Export Volume, 1000 MT*

	Export AV 93-95	GR % 87-94		
Bananas	12.465		12.465	7,29
Citrus	8.913	2,14		
Apples	4.844	4,67		
Tomatoes	3.171	4,44		
Grapes	1.968	4,70		
TOTAL VEGETABLES	15.343	5,98		
TOTAL FRUIT	36.471	5,42		

With exception of the SH, most large export forces are close to the import markets. A distinction can be made :

- Mediteranean Countries (South EU, Turkey, Israel, Maroc) exporting to the Northern European Markets. The EU imports (1996) 45% of the world trade of fruit and 44% of the world trade of vegetables. The EU is also the largest world exporter of fruit (31%) and vegetables (31%);
- Countries of Central America (Mexico, Equator, Costa Rica), exporting mainly to NA (USA and Canada) markets. In this area, the banana export occupies a major role. The USA ranks second as largest world importer of fruit (ranks second as largest world importer of fruit (19,3%) and vegetables (19,8%);
- South Asian countries (China, Thailand) mainly selling to Japan and the new emerging economies.
- SH countries, very specialised during the last 19 years. They take advantage of the off-season effect on the North European and North American Markt and are characterised by large commercial organisations.

It is clear that the overall technical progress in transport, handling and storing offered large opportunities for importing countries to diversifie their sources of import

Table: *Export Value, US $ 100.000*

	Value AV 93-95	GR % 87-94	
Citrus	4.479	-	
Bananas	4.010	10,22	
Apples	2.521	8,23	
Tomatoes	2.516	7,77	
Grapes	1.824	8,07	
TOTAL VEGETABLES	10.536 <>	10.536	10,22
TOTAL FRUIT	19.287	8,20	

Trade Policy in USA and EU

World horticultural trade is very much affected by all sorts of regulations; non-tariff barriers are complemented by a wide range of tariff barriers in thc form of export taxes and import levies. Phytosanitary restrictions can be legitimately applied to ensure food safety and to prevent the spread of diseases, but they can also be implemented for the sole purpose of acting as a barrier to trade. Most trade regulations are established to protect domestic producers from foreign competition.

However, there is a marked general trend towards liberalisation of world trade. Numerous multi-lateral and bi-lateral trade agreements have been concluded over the years. The NAFTA and MERCOSUR agreements are only two examples which have boosted fruit and vegetable trade on the American continents. Within the framework of GATT/WTO there is also a continuous move towards a further liberalisation of world trade.

The minimum access requirement, tariffication and reduction in export subsidies that was agreed upon in the latest GATT agreement, will boost fruit trade. A reduction in subsidies will disadvantage the least efficient growers. Production costs inntage the least efficient growers. Production costs in the primary sector will become the focus of competition resulting in a more significant role for global supply.

Despite these new incentives to liberalise world trade, there still is room for countries to protect their own markets and domestic producers from world market fluctuations. Even the GATT/WTO agreements have included some provisions that enable the participating countries to protect their markets. Even though all countries visibly protect their domestic produce from foreign competition, the EU is the most accused of being too protective and disruptive to the world market. The banana import regulation is the major instance of this and is treated more extensively in the sections dealing exclusively with bananas.

It is not only for bananas that the EU import regime is rather restrictive. At the beginning of 1996 the reference price system was replaced by the entry price system for a number of products. The EU replaced its fruit and vegetable reference price system by the entry price system to meet the GATT agreement which demands tariffication of import restrictions.

The entry price system is additional to a universal import duty of 8%. With the entry price system, applied to 15 fruits and vegetables, the EU sets a daily minimum import price, depending on the prices of EU produced fruit. On products entering the EU below this minimum price a progressive import tarow this minimum price a progressive import tariff is levied to equalise the import price with the minimum price. On products entering the EU at 8% or more (12% for apples and pears) below the entry price, a fixed duty is added. The entry prices are consistently set at levels well above the production costs in the EU. The entry prices vary per product and per season.

In line with the GATT/WTO treaty, liberalisation of trade is expected to continue as part of the new WTO round after the year 2000. This will result in new market opportunities, in particular for those countries with a low cost price. A reduction of subsidies will penalise the least efficient growers and different national fruit chains will witness increasing competition on production costs in the primary sector. Global supply will acquire greater significance.

Trade of Some Important Crops

Citrus

Citrus export: In 1995, 9,7% of world citrus production was exported as a fresh product, a volume of 8,9 million tons. Some 8% of oranges was exported whereas for tangerines this was 13% for lemons and for limes 14% and for the combination of grapefruit and pomelo as much as 22%. Oranges account for half of citrus exports.

The export of tangerines in particular is very concentrated with Spain alone supplying more than half of the volume of trade. Most Spanish citrus exporte than half of the volume of trade. Most Spanish citrus exports are destined for other European countries.

Spain has been able to increase exports substantially during the first half of the nineties. EU membership has enabled Greece to boost its orange exports. The decrease in Israeli exports after 1990 is said to be the result of the country's privatisation of citrus exports in 1991, the result of a hasty decision for which that the sector was poorly prepared.

Table: *Citrus Export Volume, 1.000 MT*

	Export AV 93-95	GR % 87-94	
Africa	614	-3,24	
Latin America	862	-1,12	
Asia	967	3,69	
Industrialised Countries	6.401	2,93	
EU	4.350	4,15	
NA	1.201	3,89	
TOTAL	8.913	2,14	2,14

Citrus Import: The substantial fall in orange imports in Germany and France since the late 1980s has, however, been largely compensated for by imports of other citrus fruits. The import of tangerines in particular increased but total citrus imports remained fairly stable, only France showing a slight decline since about 1990. The shift away from oranges may be induced by the demand for convenience : tangerines are more easy to peel than oranges for example. There may also have been a shift away from fresh orange consumption towards consumption of orange juice, particularly the case in Germany.

Export prices from Greece are at the lower end of the market, whereas American oranges receive premium prices.

Table: *Citrus Import Volume, 1000 MT*

	Import AV 93-95	GR % 87-94< AV>	GR % 87-94
Africa	50	13,69	
Latin America	519	13,0	
Asia	865	6,09	
Industrialised countries	5.880	0,77	
EU	4.523	0,29	
NA	579	3,58	
TOTAL	8.171	2,10	

The growth in the world citrus trade is expected to continue in the future despite the stabilisation of imports in the developed markets.

This potential for growth is due to the development and marketing of new citrus varieties which are better adjusted to consumer requirements for taste, size colour and quality. The so-called easy peelers seem to have especial potential which clearly is a result of the search for convenience.

Grapes

Grape Export: Approximately 20% of world table grape production is exported. This is very high compared to the average for all fruits of 9%. World trade in grapes, most of which are table grapes, has shown a considerable increase during the last fifteen years, Chile alone accounted for almost half of the export growth by increasing its exports from 49.600 tons (1980) to 442.820 tons. Not only Chile, but also South Africa increased exports substantially, especially after sanctions were lift in the early nineties. Specially after sanctions were lift in the early nineties.

With three countries, the EU is well represented among the major grape exporters. Most of the exports from Italy, Spain and Greece (84%, 94% and 80%, respectively in 1995) are directed toward other European countries, with Germany the major destination.

The dramatic decline of grape exports from Romania and Bulgaria, two countries that were among the major grape exporters in the early eighties should be noted. Their exports are now almost down to zero as a result of the economic problems in Eastern Europe.

Table: *Grape Export Volume* ***1000*** *MT*

	Export AV 93-95	GR % 87-94
Africa	2	36,78
Latin America	520	6,87
Asia	118	2,75
Industrialised countries	1.291	5,13
EU	920	3,72
NA	258	9,41
<8	9,41	
Economies in Transition	37	-12,2
TOTAL	1.968	4,7

Grape Import: The northern hemisphere in all absorbs more than 98% of world exports and is a major destination for southern hemisphere exports. European Union imports, approximately half of world imports, were mainly originating from other EU countries. Imports from non-EU countries also increased due to a rise in demand. The former Czechoslovakia used to be an important grape importer, but due to economic problems its position has gradually weakened since the late 1980's. The USA holds an important position both as exporter and as importer, though it is a net importer. US exports mainly Californian grapes during the second half of the year and imports are concentrated between January and May.

The growth in international trade for grapes was encouraged by the development of new varieties. The Thompson Seedless was the first seedless variety to be marketed on a large scale.

The major importers, the USA and the EU, shield their domestic producers by imposing protective measures during their own production season. The EU for instance applies its entry price system to table grape for instance applies its entry price system to table grapes from 21 July to the end of November.

Grapes imported during the EU's off-season are virtually unaffected. The USA has a similar methodology: table grape imports are subject to two marketing orders, specifying certain quality requirements, between 20 April and 15 August, in addition to the application of phytosanitary barriers.

***Table:** Grape Import, 1000 MT*

	Import AV 93-95	GR % 87-94	
Africa	3	13,55	
Latin America	76	-	
Asia	188	5,82	
Industrialised countries	1.468	2,73	
EU	929	2,45	
NA	480	3,49	
Economies in Transition	167	10,66	
TOTAL	1.922 <198>TOTAL	1.922	4,05

Bananas

Banana Export: With exports in 1995 over 13 million tonnes, almost 25% of production, bananas account for approximately one third of the world's total fruit exports. Most bananas are exported from Central America but the Philippines is also a major exporter.

The largest exporting nations have shown the largest growth during the past fifteen years, suggesting that world banana trade has become more concentrated.

Banana Import: On a world scale banana imports increased by more than 3,5 million tonnes between 1990 and 1995 despite the liberalisation in the EU's banana import regime that was implemented in 1993. With 37% of world imports, it is no surprise that the EU banana regime has given rise to so much debate.

Germany is the second largest importer after the USA, although the fall in imports since 1992 is probably due to the EU banana regime which resulted in much higher consumer prices.

Table: *Banana Export Volume 1000 MT*

	Export AV 93-95	GR % 87-94
Africa	346	9,22
Latin America	9.079	5,98
Asia	1.369	4,48
Industrialised countries	1.432	27,4
EU	1.043	44,12
NA	387/I>	
NA	387	11,85
TOTAL	12.283	7,29

Recent developments in world banana trade indicate strong growth in imports by Eastern European countries.

Table: *Banana Import, 1.000 MT*

	Import AV 93-95	GR 87-94
Africa	11	23,94
Latin America	477	11,89
Asia	885	15,86
Industrialised Countries	9.733	5,55
EU	4.599	8,86
NA	4.013	3,27
Economies in Transition	1.558	39,63
TOTAL	12.705	8,10

This is mainly the result of the banana's image of a 'luxury' fruit while still being very affordable When the economies in these countries gather more momentum, there is likely to be a slow-down as demand for other fruits increases.

China is still a relatively unknown factor as banana importer but if it continues its present trend with imports increasing more than tenfold between 1990 and 1995, from 14.000 tonnes to 160.000 tonnes.

World banana trade is likely to increase further in the future, though there are some hints that the EU market is becoming s, though there are some hints that the EU market is becoming saturated, not only because of the restrictions on imports. In developed economies bananas may have to cede market share to other newer exotic fruits.

Apples

Apple Export: The total word trade is estimated at 4,8 million tons yearly. About 1.2 million tons originates from the SH. The apple exporters not only look for increase in the existing markets but as for new markets, such as Eastern Europe and South East Asia.

It is expected that the total apple exports from the SH countries will reach 1.6 million ton by the year 2000. The SH export of Granny Smith is the most important with approximately 400.000 tons. The SH exports of Golden Delicious, Braeburn and Fuji are estimated at respectively 100.000 tons, 175.000 ton and 110.000 tons.

Particularly the USA are successful pioneers in looking for new markets in South East Asia: but Chile, New Zealand, Australia and Canada are also very active. The import of apples in Hong-Kong, Taiwan, Singapore, Thailand, Malaysia and the Philippines is increasing year by year.

The USA export to these countries is estimated at 300.000 tons in 1995 as compared to 230.000 tons in 1994. The European Union is not so active on the East Asian markets.

Table: *Apple Export Volume, x 1000 MT*

	Export AV 93-95	GR 87-94	
Africa	0,3	35,08	
Latin America	585Latin America	585	1,76
Asia	459	9,92	
Industrialised Countries	3.192	6,17	
EU	1.992	4,71	
NA	707	12,81	
Economies in Transition	608	-1,30	
TOTAL	4.844	4,67	

Apple Import: The apple import in the European Union increases year by year. The imports exceed the exports: in 1994 the EU imported 566.000 tons as compared to the export of 403.000 tons. The imports in 1995 were 693.000 ton, consisting of 590.000 ton imported from the SH (South Africa, New Zealand, Chile and Argentina.

Many countries are active on the Eastern European markets: particularly the USA and New Zealand discovered this market. A large quantity of class II apples is exported to Eastern European countries. It's is exported to Eastern European countries.

China is considered to be a potential exporter of apples. The increasing production is not only for domestic consumption but China may want to become an important supplier for South East Asia (Fuji). Of course, as the Chinese economy grows, this country is considered to be a niche market for specific varieties as well.

Table: *Apple Import Volume, x 1000 MT*

	Import AV 93-95	GR % 87-94	
Africa	51	14,15	
Latin America	389	13,64	
Asia	779	8,67	
Industrialised countries	2.653	1,16	
EU	2.359	1,43	
NA	229	-1,17	
Economies in Transition	1.085	9,5	
TOTAL	4.963TOTAL	4.963	4,53

Tomatoes

Tomato Export: Only 4% of the world production of tomatoes is exported fresh.

With approximately 700.000 tons, the Netherlands is the leading exporting country of the world. Ten percent of the export is during the winter months, coming from Canaries. Spain is the number two export country together with the Canaries providing 45% of the northern European market during the winter months.

There has been a clear increase of Spain tomato exports during the recent years.

Moroc takes advantage of the free access to the European Union. Mexico is the number three export country in the world and is almost exclusively the supplier of the USA (98% of the import of the USA). The Mexican imports cover approximately 20% of the USA fresh tomato consumption. The export of the USA comprises 225 thousand tons going mainly to Canada.

Table: *Tomato Export Volume, 1.000 MT*

	Export AV 93-95	GR % 87-94		
Africa	182	6,63		
Latin America	580	<>	580	0,95
Asia	370	2,52		
Industrialised countries	1.957	7,6		
EU	1.769	7,12		
NA	173	13,32		
Economies in Transition	83	-11,4		
TOTAL	3.171	4,44		

Tomato Import: Germany is the number one import country of tomatoes in the world. Approximately 55% of the German import originates from the Netherlands. Next is Italy and the import originating from Spain is increasing rapidly.

The number two import country in the world is the USA and the trade agreement between Mexico, USA and Canada increases trade. Surprisingly Poland and also Russia have a very strong yearly growth of import of tomatoes. The disruption of their economy and the opening of their markets gave an opportunity to Benelux exporters to increase their yearly export.

Table: *Tomato Import, 1.000 MT*

	Import AV 93-95	GR 87-94
Africa	11	15,67
Latin America	56	9,7
Asia	402	0,57
Industrialised countries	2.295	3,5
EU	1.595	4,03
NA	630	1,98
Economies in Transition	320	14,05
TOTAL	3.083	3,95

3

Plant Biotechnology and Horticultural Crops

All artificial manipulation made on a plant, or only on a part of a plant, could refer to "plant biotechnology". In our review, we shall separate the numerous plant biotechnology applications into two groups. The first group is related to tissue culture and go from a single cell, or a specific tissue, to the whole plant. These technologies are founded on a very old concept "the cellular totipotency", stated by Haberlandt (1902), but only demonstrated by Steward in 1958. His research team was the first to be able to transform a carrot cell line in some "artificial embryos, later called "somatic embryos".

The applications of the second group are more recent and derive from on excellent knowledge of the double helicoidal DNA structure proposed in 1953, by James Watson and Francis Crick. They base their strategy on the molecular biology science.

Horticultural Biotechnologies Related to Tissue Culture

The development of plant biotechnology has been related to the discovery of the growth regulators and their progressive introduction in tissue culture. In 1934, in the States, white was producing continuous growth of in vitro tomato roots. After that, went and Thiman's work on auxin, Gautheret in France (1935) and white in US (1938) were auxin, Gautheret in France (1935) and white in US (1938) were able to obtain division of isolated cells from Salix and tobacco, and finally to maintain undefined cell lines.

In 1957, Skoog and Miller demonstrated that 6-furfurylaminopurine, a cytokinin isolated from a DNA hydrolysate, could induce some caulogenesis, but also that the cytokinin/auxin balance could start different morphological programmes. A ratio higher than one unit induced caulogenesis, a ratio inferior at this unit induced roots. On this basis, a multibillion dollar industry was created. Hundreds of small and large nurseries and biotech labs throughout the world are propagating in vitro more than 1000 different plant species (Bajaj, 1991).

This micropropagation technique offers not only means for mass propagation, but also plays an important role to conserve elite or rare plants that are threatened with extinction. Moreover, meristem tip culture offers the possibility of virus eradication. These technologies started about 20 to 25 years ago, especially for ornamental plants, and for a few other crops: potato, strawberry, bananas... Propagated clonally, it was very important to have a true to type propagation technique.

On the other hand, it's also possible to use tissue or cell culture to increase genetic variability. Undifferentiated cells obtained from callus, cells or protoplasts culture are produced and submitted to an selective pressure to improve a and submitted to an selective pressure to improve and fix the somaclonal variants.

Production of True to Type Plant

Meristem Tip Culture

Production of Virus-free Plants: In 1952 Morel and Martin were successful in regenerating a virus-free dahlia plant by the excision of some meristematic domes from virus infected shoots. Three years later, the same authors were able to eliminate virus A and Y from virus infected potato. Semal and Lepoivre (1992) reported that a virus-free sweet potato was producing 40T/ha in China by comparison of the 20 T/ha produced before meristem culture.

Virus eradication is dependent on several parameters. But to take advantage of the non uniform and imperfect virus distribution in the host plant body, the size of the excised meristem should be as small as possible. For Stone (1963), only tips between 0.2 and 0.5 mm most frequently produce virus free carnation plants. The explants smaller than 0.2 mm can't survive, and those larger than 0.7 produce plants that still contain mottle virus.

Today, no definitive explanation can be given to understand this virus eradication (Wang & Charles, 1991). various explanations have been given: absence of plasmodesm in the meristematic domes, competition between synthesis of nucleoproteins for cellular division and viral replication, inhibitor substances, absence of enzymes preiral replication, inhibitor substances, absence of enzymes present only in the cells of the meristematic zones, and suppression by excision of small meristematic domes. This last proposal could explain why some potato plant showing virus particles in the meristematic domes, could regenerate a virus free plant (Mellor & Stace Smith, 1977).

Shoot Tip Micrografting

When meristem tip culture fails, it is possible to graft small meristematic domes on young seedlings growing in vitro. In this way, Navarro et al. (1975) eradicated all the virus diseases from Spanish Citrus orchards. This technique was also very successful in eliminating virus diseases from peach trees (Mosella et al 1980).

Factors Influencing Virus Elimination

The success in virus elimination depends on the choice of the explants, the virus, temperature,....In general a high temperature (37C) is effective for virus diseases, but is depressive for viroids. The introduction of virus inhibitors (ribavirin) into the culture medium can't inactivate the viruses, but it prevents their replication. A combination of cytokinin and ribavirin, or a heat treatment in vitro, are sometimes used with success to inactivate some viruses difficult to eliminate by meristem culture (NRSV, PDV,..).

Micropropagation Techniques

During the Micropropagation Process, the Genetic Stabilit

During the micropropagation process, the genetic stability of the new shoots is dependent upon their origin. Axillary shoots issue from pre-existing buds and are normally true to type, indeed the meristematic cells are genetically very stable.

Adventitious shoots, such as somatic embryos, are neo-formed buds developed directly on some organs, or indirectly through a callus phase formed on this organ. So, if the mother plant presents a cell mosaic or chimaeric tissues, risks of genetic variation exists. It is similar in the case of an indirect regeneration, when the callus phase is too long.

Propagation by Axillary Shooting

This technique has proved to be the most applicable and reliable method of in vitro propagation. Axillary shoot growth is stimulated by overcoming apical meristem dominance.

Commercial tissue culture laboratories are now able to propagate a large number of herbaceous ornamental species and several woody plants in this way. However, the propagation of Pelargonium, Howea, and a few other horticultural plants are always difficult to propagate by axillary branching.

Propagation by Direct or Indirect Organogenesis

Adventitious shoots could arise directly from the tissue of explants without callus formation. Several plants of the family gesneriaceae (Saintpaulia, Streptocarpus,...) regenerate directly buds on leaf explants, likewise Lilium rnerate directly buds on leaf explants, likewise Lilium regenerates on scales.

However more often, like for Ficus lyrata, adventitious buds appear on callus. While coffee, cocoa trees, and many conifers are produced by somatic embryogenesis developed on callus or cell suspensions.

Constraints of in Vitro Micropropagation

The establishment of axenic culture could be difficult when the explants are coming from hot and humid countries.

Moreover, in the course of micropropagation several authors observed a sudden appearance of endogeneous bacteria (Cassels, 1997). Nevertheless only Holland and Pollaco (1994) have demonstrated the presence of this kind of bacteria in more than 70 different species. For Leifert and Woodward (1997), the major source of contamination is the initial explant. They add that microbial contamination in commercial plant tissue culture laboratories is the most important cause of losses. For this reason, they support the introduction of a microbiological production control strategy.

A second important constraint during the initial phase mainly concerns woody plants. A mature tree must be rejuvenated to recover its morphogenetic competence. Franclet (1981) described various treatments, micrografts, cascade grafts, cytokinin treatments to reverse the adult phase to juvenile. The micropropagation of Acacia Senegal was only possible after rejuvenation by micrografting (Palmnly possible after rejuvenation by micrografting (Palma et al. 1997).

A distinction between chronological, ontogenic and physiological age (Monteuuis 1989) is needed to understand the rejuvenation concept. This was definitively demonstrated after a complete reversion of Sequoia sempervivens (Franclet & Franclet Mirvaux, 1992).

Improvement of Axillary Branching

The cost of micropropagated plantlets is also an important limitation of the techniques. In New Zealand, where they produce 2-3 million micropropagated radiata pine per annum, the relative cost of micropropagated planting stock had dropped from 13,8 times the cost of seedlings in 1988 to 6,9 by 1993 (Smith, 1997).

To reduce manpower costs, several improvements have been proposed. The more simple method was in vitro layering developed by Wang (1977) to clone PVX-free potato plants. The first plantlets placed on the medium in a horizontal position developed axillary shoots. They are harvested by cutting one centimetre above the medium surface, at 3 weeks intervals. A similar technique called 'hedging system ' by Aitken Christie and Jones (1987) was later used to produce Pinus radiata.

Ziv (1990) proposed for corm plants, Gladiolus and Nerine, a very rapid propagation system. She reduces the internodes and leaves by introduction of an anti-gibberellin agent in the medium. Finally, only aggregates of buds are formed, then they finally, only aggregates of buds are formed, then they are divided and introduced in bioreactors for mass production. Similar systems were developed in Gembloux, to propagate carub trees and asparagus.

Since 1988, Duhem was producing very large quantities of Eucalyptus plantlets in Petri dishes without anti-gibberellin but in complete darkness. Transfers from one Petri to another is made by a simple squashing (Boxus et al, 1991).

Somatic Embryogenesis Propagation

For genetically stable species, somatic embryogenesis offers a very fast scaling-up system, especially when it's possible to produce embryos in bioreactors. Unfortunately this production via fermentors was not as simple as first envisaged. And today, only a few model plants are successfully produced by such technology: carrot, celery. Other applications remain at the experimental stage: coffee, oil, palms, conifers, Euphorbia pulcherrima,... and several other horticultural species.

Several bottlenecks limit the use of this interesting technology. One of the main problems is genetic stability. So, despite the clonal nature of nucellar embryos, different morphological anomalies can occur among mango somatic embryos, as it was also observed in Citrus plants derived from nucellar cultures (Litz et al., 1993).

Another difficulty is the loss of embryogenic capacity over time, a phenomenon observed with different species. It is also a phenomenon observed with different species. It is also important to note that somatic embryogenic lines of conifers are always originated from immature embryos.

Artificial Seeds

Another very interesting possibility of the somatic embryogenesis technology has been developed during the past fifteen years by Redenbaugh and his team (1991, 1993). They were able to encapsulate somatic embryos by hydrogel coatings (sodium alginate), producing single embryo artificial seeds. To date, some improvements offer the possibility to directly plant the artificial seeds in the greenhouse on special substrates (vermiculite, sans,...). This methodology will provide in the future a good technique to reduce the cost of transplants.

Gene Bank

Tissue culture methods offer the opportunity for in vitro collecting, rapid multiplication and distribution of important, elite, or rare plants that are threatened with extinction. The two major in vitro storage strategies are slow growth and cryopreservation. Since the first results of Seibert (1976), who was able to initiate shoots from carnation shoot apices frozen to -196C, this technique is now successful for many of horticultural species. Dereuddre et al. (1991) have provided a very simple technology to freeze encapsulated meristems in dried alginate beads. It works for pear, strawberry, eucalyptus, potato,....

Some lor Pear, Strawberry, Eucalyptus, Potato

Some international institutions have very large collections of old and current varieties, available for exchange and introduction in crop improvement programmes. The International Potato Centre (CIP) in Lima, Peru, has a large world potato collection. Germplasm of sweet potato and cassava is at the International Institute of Tropical Agriculture (IITA), Ibadan, Nigeria.

The movement of germplasm involves the risks of accidentally introducing plant quarantine pests along with the host plant material. To limit these risks, the plant material should be transferred from one country to another as in vitro cultures through a transit Centre, where it should be indexed. For bananas, in the framework of INIBAP, the transit Centre is the Catholic University of Leuven in Belgium where a very large in vitro germplasm exists.

Tissue Culture Techniques

Somaclonal Variations

The production of plantlets by callus regeneration, cell suspensions, protoplast cultures, could present some deviations with regard to the mother plant. This is a way to increase the genetic variability. Associated with a selective pressure (stress to toxins, pH, salinity, cold,...) it's used to obtain resistant lines. Indeed, after regeneration, plants can express new potentialities rarely obtained another way. Stable and profitable variants are sale another way. Stable and profitable variants are selected and introduced in breeding programmes. In 1976, a Pelargonium cv Velvet Rose was created by this technique (Reisch, 1983). Later, other applications were described, as in sugarcane, tomato, red pepper,... Following Sibi (1994), extrachromosomic systems must be involved to explain the genetic behaviour of some new expressed tissue culture characteristics. She uses the term "epigenic" to evoke a whole of hereditary elements which belongs to the cytoplasmic or nuclear compartments, but are not transmitted by mendelian rules.

Haplodiploidisation

Many species are able to produce haploids through different in vitro techniques. The oldest was anther culture of androgenesis. To date, the results vary considerably from one species to another. Solanae (datura, tobacco, red pepper, eggplant, petunia), cereals (wheat, barley, rice, triticale, maize) or crucifers (soybean, cabbage) are species easy to regenerate by anther culture. To the contrary, tomatoes, leguminous, or Compositae are recalcitrant.

Ovule culture, or gynogenesis, was a successful technique for Gerbera, beet, courgette,... However, the most common technique is pollination with irradiated pollen, in order to induce in vivo parthenogenesis. The oospheres developed in embryos without fertilisation are saved by embryo rescue.

The haploid plantlets (n chromosomes) are tree. The haploid plantlets (n chromosomes) are treated with colchicine to produce fertile homozygous lines, called "double haploid lines". This haplodiploidisation technique could give immediately new elite genotypes, hybrid parents after in vitro propagation, as asparagus supermales (MM), or useful genetic material to establish gene mapping. A good synthesis book concerning all these possibilities was published by Bajaj (1990).

Protoplast Culture

Protoplasts are the smallest units able to regenerate a whole plant. Therefore protoplasts cultures can serve to enlarge genetic variability by introducing somaclonal variation. However, the main interest of protoplasts, is their capacity to fuse and to produce hybrids or cybrids, new organisms most often unknown in nature. Indeed, protoplasts allow to transgress the barrier of the botanical species or genus. Naked protoplasts can accept without rejection external elements: nuclei, cytoplasmic corganelles, liposomes,.... containing genetic information. The first protoplast fusion application was a cytoplasmic transfer from one genotype to another to induce male sterility (CMS) from mitochondrial origin. These male sterile hybrids are interesting to produce F1 hybrids (Brassica, Cichorium,...).

Another cytoplasmic transfer concerns the introduction of chloroplastic DNA to induce herbicide (atrazine) resistance, from Solanum nigrun duce herbicide (atrazine) resistance, from Solanum nigrun to tomato for instance (Jain et al. 1988).

Many intergeneric and interspecific hybridations are possible but not always profitable because these hybrids are too often infertile. This problem does not exist if we can vegetatively propagate the hybrids, as in the case of Citrus rootstocks. In Florida, Grosser (1990) was able to analyse several thousand Citrus hybrids.

Contributions of Molecular Biology in Horticulture

More and more pathologists are using molecular probes for early detection of diseases. However, the main contributions of modern biotechnology remain probably in the hands of the plant breeders. Indeed, molecular markers are of great use in detecting desirable new genes, or to identify important QTL. This marker-assisted selection is very useful for many vegetable plants. Other applications are DNA - fingerprinting and genetic engineering. This recent technology aims to insert and to express new specific genes into a selected plants.

Marker-assisted Selection

Molecular markers and genome mapping will reach a large extension in the next breeding programmes of tomatoes, peas, cabbages, melons, potatoes,...

It will be possible to improve the speed and the incorporation efficiency of desirable new genes by utilizing of closely linked selectable molecular markers genes by utilizing of closely linked selectable molecular markers. The number of backcross will be reduced. Tanksley et al. (1981) show that only with 12 markers, one on each tomato chromosome, the composition of the recurrent parent genome is similar to that observed in the absence of selection after the third backcross two years later. Therefore, it's easy to understand the high improvement that we can get with the 700 tomato markers already known in 1990 for this crop (De Verna and Paterson, 1990).

The utilization of molecular markers could also identify important quantitative trait loci (QTL). Recently Foolad and Chen (1998) have identified 13 RAPD markers at eight genomic regions, that were associated with QTLs affecting salt tolerance during germination in tomato.

DNA Fingerprinting

The DNA polymorphism observed by RFLP, AFLP, or RAPD, allows cultivar identification of fruit trees, apple, citrus,... or other vegetatively propagated plants. Although some mutants, as coloured fruits, escape these identification techniques, some people would evaluate the genetic uniformity of regenerated plants in this way.

Genetic Engineering

Genetic engineering consists of introducing a foreign gene into a plant genome to create a new function. Or at the contrary, it could be to reduce or suppress an existing gene function, as in the antisense strategy.

1. *Historic:* In this field, progress were very rapid. In 1974 the pathogenicity of Agrobacterium tumefaciens, due to a large plasmid called Ti, was demonstrated. In 1977, Chilton et al. demonstrate a stable incorporation of plasmid Ti-DNA into higher plant cells. In 1983, the team of Van Montagu and Schell in Gent succeeded in producing a transgenic plant, and eleven years later the first transgenic cultivars were available in the market.
2. *Transgenic Plants with New Agronomic Traits:* Two different techniques are commonly used to introduce genetic information (DNA) inside cells, protected by pectocellulosic walls. A particle bombardment process, called biolistic, is used with success for monocots gynmosperm, squash, peas,...Today, this technique allows the bombardment of meristems and avoids the difficulties of regeneration. But the mediation of Agrobacterium remains till now the most routinely system to transform dicotyledonous plants.

The main agronomic traits introduced in horticultural plants and already commercialised are Bt toxin and herbicide resistance. Other studies concern virus resistance, male sterility...

3. *Antisense Strategy:* Calgene created in 1994 the first commercial transgenic plant, a long shelf life tomato, by the suppression of polygalacturonase activity due to an antisense gene (Smith, 1988). However, t activity due to an antisense gene (Smith, 1988). However, this Flavr Savr tomato variety was removed from the trade 3 years later, because of its disease susceptibility and its lack of productivity !

   Later, other tomato varieties with long storage qualities were obtained by the utilization of an antisense RNA inhibition of ACC synthase or ACC oxydase, two ethylene precursors.

   The antisense technique was also used to reduce the lignification of woody plants, by blocking the enzymes involved in the precursor of lignin biosynthesis. Another interesting application was the induction of white flowers in petunia and in different other ornamental plants by the suppression of chalcone synthase activity.

4. *Transmission of the New Traits:* These traits induced by the transferred DNA are transmitted to the progeny as a dominant Mendelian character. Nevertheless in some cases, transgenic plants with a strong gene expression can generate progenies with only a faint no gene expression. This problem is not clearly elucidate.

5. *Marker, Reporter, Promotor and Expression Genes:* To select the transformed cells, some marker genes, very often resistant to antibiotics or to herbicides, are attached to the coding sequence, as a promoter or an expression gene. Those allow the new gene to express in the whole plant or in a specific plant tissue.

   Some reporter genes are used to follow the evolution one.

   Some reporter genes are used to follow the evolution of the transformed cells. The most frequent reporter genes are the gene "gus" or the gene lux (luciferase).

6. *Horticultural Transgenic Crops already released in USA:* The Flavr Savr tomato was the first genetically engineered whole food approved for commercial sale in 1994. Four other transgenic tomatoes with a delayed ripening were approved later (1995 and 1996).

In 1995, potatoes with Bt genes and a squash cultivar resistant to two viruses were released. Two years later, another squash cultivar resistant to 3 viruses and a papaya line also resistant to viruses were approved. Two other horticultural crops are waiting the authorities approval: red hearted cichory (Radicchio) with male sterility and resistance to herbicide, and a tomato with Bt toxin.

It there are so few commercial transgenic plants in horticulture, it's probably due to the relatively low acreage of these horticultural crops in comparison with other agricultural crops. In 1997, in the world 5 million ha were planted with glyphosate resistant soybeans released by Monsanto alone.

Nowadays, it's true also that many transgenic plants exist in research laboratories, awaiting authorization for field testing, as they are very interesting model plants for learning physiology: lettuce with less nitrate by increasing nitrate reductase gene active with less nitrate by increasing nitrate reductase gene activity, a yeast ribonuclease gene to reduce viroid infection on potato, modification of lignin synthesis in plant to produce trees adapted to paper industry, or timber, or biomases production, regulation genes to control tree architecture,....

Situation of Biotechnologies in the World: Development and Risks

Situation in the Young Countries or in Developing Countries: A general use of in vitro micropropagation in the developing countries is now a reality, as it was predicted by Albert Sasson (1993). African countries for instance are producing potatoe, banana, cassava...

Even the production of transgenic plants is no longer a prerogative of the Northern countries. Different experiments are now starting in the South, specially in the international laboratories from CGIAR. Although CGIAR's expenditures on biotechnology for 1997 raise only 24.2 million US$ on an annual budget of 345 million US$ (Biotechnology and Development Monitor, December 1997, 33, pp12-17) this Consultative Group on International Agricultural Research plays a very important role as contributor to agricultural research for developing countries.

The large international research centres IITA (Ibadan, Nigeria), CIP (Lima, Peru), CIAT (Cali, Columbia),... belong to the CGIAR. Their roles and NGO's are primordial. The genetic transformation their roles and NGO's are primordial. The genetic transformation of cassava is an excellent example. Its tuberous roots provide food for over 500 million people, mostly small-scale farmers. Unfortunately, till recently this integral plant for food security in developing countries has been recalcitrant to transformation approaches. However, thanks to the participation of international institutes located in cassava growing countries, CIAT and IITA with the help of four development associations from Swiss, UK, The Netherlands and USA, were able to initiate procotols for cassava transformation and regeneration, (Biotechnology and Development Monitor, March 1997, 30, pp. 16-18).

Other networks, like the Cassava Biotechnology Network (CBN) previously described, exist. REDBIO, a technical cooperation network on plant biotechnology, is supported by FAO to promote a best use of scarce manpower, equipment and other resources in Latin America.

On the other hand, more and more private laboratories exist in the developing countries which have a high scientific and technical level (China, Singapore, Taiwan, India, Brasil, Mexico, Chile,...). In these countries, European, American or Japanese agrochemical companies are investing. Also these countries will have their chance to develop their own technologies, and to export to the Northern markets.

Meanwhile, genetic engineering techniques are being applied mainly to crops which are important for g applied mainly to crops which are important for the industrialized world, not crops on which the world's hungry depend. Therefore, it is unrealistic when Monsanto writes "the experts said that biotechnological innovation will increase the crop productivity without occupation of new lands, saving tropical forest of best quality and animal habitat".

However, the World Bank was promising in a 1997 publication Bioengineering of crops that" transgenic crops could improve food yield by up to 25 per cent in the developing countries and could help to feed an estimated additional three billion people over the next 30 years".

But, not all people are confident in this prospective. To the contrary, some are afraid that the poorest countries, where more than 700 million people are chronically undernourished will miss the biotechnological revolution.

Nevertheless, there is some hope to transfer to the developing countries high technology, which don't undermine the environmental and social network. So, the Agricultural Biotechnology for Sustainable Productivity Project (ABSP project) is managed by Michigan University. Four American universities, two research centres from

France and USA, and two American private companies are also involved in this project. The ABSP's objectives are the reduction of losses due to pathogens and the reduction of losses due to pathogens and pests by using transgenic plants (sweet potatoes, potatoes, cucurbits, tomatoes) and by cloning commercial value-plants (bananas, pineapple, coffee).

Situation in the Northern Countries

1. *Mass Propagation:* In 1988, Pierik (1991) predicted a European in vitro propagation of 212 million plantlets/year. He was also predicting an important development of this activity for the future. Following Pierik this quantity accounted for only 5 % of the plants able to be propagated by vegetative means. However a few years later Pierik's prediction was not confirmed. The world production was estimated at 600 million. In USA, Zimmerman (1997) was recording a production of 121 million plantlets for a total of 110 laboratories, half produced in 11 laboratories, and 6 labs were propagating each more than 6 million plants per year.

   In Europe, only the Dutch production of micropropagated plants is precisely known. In 1990 the total production was 95.5 million. In 1995, the Dutch in country production reached only 53,7 million, whereas importations increased 77,3 million. (37 million from Poland and 17,1 million from India) (Pierik, data unpublished). These laboratory relocations to low-cost manpower countries are commonly observed, and put the question of the quality management.

2. *Transgenics:* The GMOs invade progressively all the US lands: 1.6 % in 1996, 15 % in 1997, 50 % in 2000..... It will represent an estimated market of 3,9 million US$ in 2003 ! The largestt agrochemical companies are merging to become giants of agrobusiness such as Novartis, Monsanto, Zeneca,...

   Following Monsanto in 1996, about 950 T insecticides were saved by introduction of resistant cotton, id est a net savings of 81 US$/ha for the grower. Glyphosate-resistant soybeans (4 million ha in 1997) allow a savomgs of 44 to 49 US$/ha. (Information from cultivar, suppl. n 436, Feb. 16, pp. 24-37, 1998).

3. *Perspectives, Limitations and Environmental Risks*: Ecological impact related with the introduction of GMOs is always an open debate.

The application to agriculture of these new technologies certainly opens interesting perspectives, but also raises potential problems. The risk of crop transgene spreading has been demonstrated. A researcher of Clemson University in South Carolina reported "that in a population of wild strawberries growing within 50 meters of a strawberry field, more than 50 % of the wild plants contained marker genes from the cultivated strawberries" (Kling J. 1996).

A Danish team (Mikkelsen et al, 1996) have shown a possible rapid spread of genes from oilseed rape to the weedy relative *Brassica campestris.*

There are other risks. *Brassica campestris.*

There are other risks. The introduction of Bt gene allows a drastic reduction in the use of toxic chemicals for crop protection. But a poorly controlled use of Bt-technology can destroy more effectively the predators than the pests.

Or when" many crop plants are transformed with similar effective traits, in such situation, many polyphagous pest species, which by nature are more flexible evolutionarily than those that have a narrower diet, are likely to overcome Bt resistance very quickly". (Hokkanen, 1998).

Therefore, before releasing a transgenic plant, any risk has to be weighed against the benefit of the transgenic crops. We must not forget that annually in the world 500.000 acute pesticide poisonings, with 5000 deaths, are observed.

Table: *Estimated Global Economic, Environmental, and Human Health Benefits of Bt Transgenic Plants. (following Hokkanen, 1998)*

Global Market Penetration of Transgenic Plants (% use)	Reduced Use of Conventional Insecticides (US$ millions)	Cost of Transgenic Plant (US$ millioTH=112> Cost of Transgenic Plant (US$ millions)	Economic Gain (US$ millions)	Estimated Environmental Gain (US$ millions)	Estimated Human Health Gain (US$ millions)
1	90	45	45	91	16
10	900	47	853	917	157
25	2250	49	2201	2293	394
50	4500	51	4449	4585	787

Horticulture Research in India? Infrastructure, Achievements, Impact, Needs and Expectations

India has a wide variety of climate and soil on which a large range of horticultural crops such as, fruits; vegetables, potato and other tropical tuber crops; ornamental, medicinal and aromatic plants; plantation crops; spices, cashew and cocoa are grown. After attaining independence in 1947, major emphasis was laid on achieving self sufficiency in food production. Development of high yielding wheat varieties and high production technologies and their adoption in areas of assured irrigation paved the way towards food security ushering in green revolution in the sixties. It, however, gradually became clear that horticultural crops for which the Indian topography and agro climates are well suited is an ideal method of achieving sustainability of small holdings, increasing employment, improving environment, providing an enormous export potential and above all achieving nutritional security. As a result, due emphasis on diversification to horticultural crops was given only during the last one decade.

Research Infrastructure

The Indian Council of Agricultural Research is the premier agency which pioneered systematic research on agricultural crops in the country. Horticulture research in India received very little attention till the 3rd Five Year Plan. The establishment of the Indian Institute of Horticultural Research at Bangalore and starting of eight All India Coordinated Crop Improvement Projects to cover different horticultural crops was a landmark in the history of horticulture in 4th Five Year Plan (1969-74). Rapid expansion of infrastructure took place in 7th and 8th Plans. Today, the horticultural research in the country is being carried out at eight ICAR institutes (with 26 regional stations), 10 National Research Centres (on major crops) and a Project Directorate on Vegetable crops.

Area specific, multi-disciplinary research is also being conducted under 14 All India Coordinated Research Projects each on Tropical, Subtropical and Arid Fruits; Vegetables, Potato, Tuber Crops and Mushrooms; Ornamental Crops, Medicinal and Aromatic crops; Palms, Cashew, Spices and Betel vine; and Post Harvest Technology at 215 centres located at various research Institutes, and State Agricultural Universities. In addition, four net work projects each on hybrid research in vegetable crops, drip, irrigation in perennial horticultural crops, protected cultivation of ornamental crops and *Phytophthora* diseases of horticulture crops are now in operation. Research on horticulture is also being undertaken at several multi-crop, multi-disciplinary Institutes. Departments of Horticulture in 24 Agricultural Universities, one deemed to be University and one full fledged University of

Horticulture and Forestry are also engaged in horticultural research. Besides 280 adhoc schemes supported from Agriculture Produce Cess Fund and a number of foreign-aided projects have also been in operation on specific problems of different horticulture crops. As a result, the country now has a sound research infrastructure in horticulture to meet the growing needs and expectations of the fast developing horticulture industry.

Budgetary Support

The investment in horticulture research by the ICAR in the Central sector has increased significantly in the last two Plans. The Plan allocation for horticultural crops started in 4th Plan (1969-74) with a modest allocation of Rs. 34.78 million and was enhanced to Rs. 319.56 million in the 7th Plan (1985-90) and to Rs. 1047 million in the 8th Plan (1992-97). Non-Plan expenditure also increased from Rs. 73.55 million in the 5th Plan to Rs. 768 million in 8th Plan. Overall increase in Plan investment in 25 years has been of the order of 2775.21 per cent. The per cent budget allocation for horticulture research out of the total budget for agriculture research rose from 6.1 in 5th to 6.5, 6.67 and 7.7 in 6th, 7th and 8th five year plans, respectively. Similarly, expenditure for Central Sector Schemes of the Department of Agriculture & Cooperation for horticulture crop development also rose tremendously from Rs. 20.5 million (4th Plan) to Rs. 76.18 million (5th Plan), Rs. 146.37 million (6th Plan), Rs. 250 million (7th Plan) and Rs 10,000 million (8th Five Years Plan).

Table: *Budgetary Support (Million Rupees)*

Five year plans	*Total for Agriculture Research*	*Share of Horticulture (%)*	*Total Development Support*
IV 1969-74	6105	N.A	20.50
V 1974-78	7292	6.10	76.18
VI 1980-85	10684	6.50	146.37
VII 1985-90	8445	6.67	250.00
VIII 1992-97	15165	7.70	10,000.00

Manpower

Nearly one sixth of the total strength of 5906 scientists working in ICAR is allocated for horticulture research in ICAR Institutes. Besides, approximately, 560 scientists are working in State Agricultural Universities in ICAR funded All India Coordinated Projects. In addition, a large number of scientists are working on horticultural crops in State Agricultural Universities.

Total Scientists in ICAR	5906
Scientists working in the Division of Horticulture	832
Scientists working in other ICAR Institutes	150
% of Total	15
Scientists working in SAUs in ICAR funded projects	560

Research Achievements

Introduction and Cultivation of New Crops: Several new crops have been introduced for commercial cultivation, *e.g.*:

- Kiwi fruit in sub-mountain areas of North India
- Olive in mid hills of North Western Himalayas
- Low chilling stone fruits in the North Western plains
- Oilpalm in coastal states of Karnataka, Andhra Pradesh, *etc.*
- Gherkin in south and west India
- Baby corn and sweet corn in certain specific pockets, and
- Broccoli, Brussels' sprouts, asparagus, celery, parsley near the cities.

Crop Improvement

- A large number of high yielding varieties developed in several horticultural crops *e.g.* fruits (76), vegetables (160), potato (29), other tuber crops (24), ornamental crops (300), palms (20), spices (51), cashew (33) and betel vine (1).
- First seedless variety of mango developed.
- 40 F_1 hybrids developed in brinjal, tomato, chillies, cauliflower, carrot, capsicum and muskmelon.
- Self incompatible lines in cauliflower, gynodioecious lines in cucumber and muskmelon, genetic male sterile lines in tomato and temperature tolerant strains of button mushroom developed.

Propagation of Quality Planting Material

- Standardized propagation technique for many fruits hitherto propagated by seed. *e.g., aonla, bael, ber,* black pepper, cardamom, cashew, cassia, cinnamon, clove, custard apple, jack fruit, *jamun,* nutmeg, sapota and walnut.
- Standardized Seed Plot Technique resulting in successful disease free potato seed production in the tropics and sub tropics of the country. Standardized method of micro-propagation and *in vitro* micro-tuber production in potato.

- Identification of suitable parental lines for production of True Potato Seed (TPS) and Standardized technology for raising commercial crops.
- Micropropagation protocols developed in banana, black pepper, betel vine, cardamom, ginger and turmeric.
- Production of coconut hybrids through establishment of Seed gardens of Tall (T) x Dwarf (D) and D x T hybrids.
- Standardized rootstocks in citrus, grape and apple.

Agrotechniques

- Standardized high density plantations in banana, citrus, mango and pineapple and high production technology in several crops *e.g.*, pineapple, black pepper and cardamom.
- Year round production technology in tomato and "off season" cultivation of onion and cauliflower developed.
- Arecanut, coconut and potato based cropping systems developed to maximize productivity under high management conditions.
- Standardized use of several plant growth regulators and chemicals now commercially employed in production and quality improvement of horticultural crops *e.g.*, paclobutrazol for induction of flowering in mango; gibberellic acid for improving berry size and quality in grape; Maleic hydrazide for preventing sprouting in onion and potato; Dormex for hastening bud burst in grapes; and Boron and Calcium for changing flower cycle in some cucurbits.

Crop Protection

- Developed improved disease detection techniques such as ELISA and ISEM for improving seed quality and tissue culture technique for rapid multiplication of potato.
- Developed IPM for fruit borer in brinjal, diamond back moth in cabbage, thrips in chillies, phytophthora foot rot in black pepper, "Katte", rhizome rot in cardamom, rhizome rot in ginger, late blight and bacterial wilt in potato.
- Developed biological control measures for mealy bug in grape, fruit borer in tomato and okra.

Post Harvest Management

- Preharvest treatments to control post harvest losses in citrus, mango and grape standardized.

- Maturity standards for mango, guava, grape, litchi and ber standardized.
- Chemical treatment for regulation of ripening in mango, sapota and banana standardized.
- Optimum storage temperatures worked out for several fruits, vegetables and tuber crops.
- A mango harvester, fruit peeler, hand and pedal operator cassava chipping machine, harvesting tools (5-14 times efficient); Implements for mechanization of potato cultivation *e.g.,* oscillating tray type potato grader, fertilizer application cum line marker, potato culti-ridger, soil crust breakers, potato digger and automatic potato planter/diggers developed.
- Low cost environment friendly storage system for fruits, vegetables, potato and onion developed.

Impact of Research

Fruits: Area under fruits increased from 1.22 million hectares to 3.35 million hectares in 1995-96. India with a production of 41.50 million tonnes (1996-97) is the second largest fruit producer next to China (45.46 million tonnes) with a share of 8% in world fruit production. India produces 65% and 11% of worlds mango and banana, respectively, ranking first in the production of both the crops. It has the highest productivity in grape in the world. Significant expansion has taken up in apple*, aonla, ber,* pomegranate and sapota cultivation.

Vegetables: India ranks second in the world vegetable production after China (71.59 million tonnes). Vegetable production has increased three times during the last 50 years. A large area is now covered with F_1 hybrids in vegetable crops resulting in increased yield and better socio-economic status of farmers. Vegetables like tomato, cabbage, cauliflower, radish and onion are now produced almost round the year. India has attained self sufficiency in seed production of temperate vegetables.

Mushroom: Mushroom cultivation has spread to almost all parts of the country. Its production has increased from 100 tonnes in 1970 to 30,000 tonnes in 1996-97. The productivity has increased from 10-12 kg m^{-2} in 1985 to 18-22 kg m^{-2} in 1995.

Potato: In potato, area, production and productivity has increased from 0.234 million hectares, 154 million tonnes and 6.59 t/ha in 1949-50 to 1.14 million hectares, 1924 million tonnes and 16.9 t/ha respectively. This increase is 12, 4.9 and 2.6 times, respectively. The annual compound growth rate for potato during this period was 6.07 compared to 5.6% for wheat, 2.7% for rice and 2.74 % for total food grains. India is the only country in South East Asia having a national disease free seed production programme producing 2600 tonnes of breeder's seed annually.

Table: *Production, Demand and Projections of Horticultural Crops*

	1996-97		Demand		Target 2002	
	Area*	Production**	Area	Production	Area	Production
Fruits	4.54	46.97	5.24	59.47	4.93	56.00
Vegetables	5.12	80.80	6.96	131.2	5.73	108.00
Spices	2.54	2.78	2.94	4.43	2.60	3.90
Coconut	2.00	9.75	2.67	15.60	2.57	15.00
Cashew	0.63	6.45	0.69	0.80	0.60	0.70
Arecanut	0.24	0.31	0.24	0.39	0.25	0.39
Total	15.07	141.06	18.74	211.89	16.68	183.99

**Area in million hectares;*
***Production in million tonnes*

Cassava: In cassava, productivity has increased from 7t/ha in 1960-61 to 22t/ha during 1992 which is more than double the world average (9.81t/ha). Sago and starch industry based on cassava has been developed.

Floriculture: Protected cultivation of cut flowers started a decade back and India has already entered the world cut flower market. Micro propagated ornamental foliage plants are being exported in millions internationally. Export of dried flowers from India is increasing.

Coconut: In coconut, area has increased from 1 million hectares in 1980 to 1.793 million hectares in 1996-97. India has become one of the largest coconut producing country of the world. Production of coconut has gone up from 5677 million nuts to 13968 million nuts. The productivity in coconut has increased from 5249 nuts/ha to 7808 nuts/ha. Coconut contributes 700 billion rupees to the GDP of the country. The contribution of the crop to the total edible oil pool in India is around 6 per cent. India also exports coir and coir products derived from coconut husk to the tune of 2260 million Rupees.

Export of Horticultural Produce

Arecanut: India continues to dominate the world in area, production and productivity of arecanut and has achieved self sufficiency in arecanut production (0.27 million tonnes). Most of the production is domestically consumed.

Oilpalm: Area under oil palm has gone up from 200 ha in 1965 to 40,700 ha in 1996-97. Average productivity in oil palm plantations is now 4-5 tonnes/ha which compares favourably with other countries.

Spices: India is the largest producer (2.48 million tonnes), exporter (0.20 million tonnes) and consumer of spices. Indian spices flavour foods in over 134 countries. Spice exports touched Rs. 11800 million during the last year.

Cashewnut: In cashewnut, area has increased from 0.176 million hectares in 1961 to 0.659 million hectares in 1996-97. The production in cashew has gone up from 0.079 million tonnes to 0.430 million tonnes in 1996-97. India exported cashew kernels worth Rs. 13000 million (362 million $US during 1996-97). Export of cashew rising @ 27% per annum. These export earnings are exceeded only by coffee and rice among agri-exports.

Needs and Expectations

In spite of significant achievements in horticulture R&D, a number of challenges still need to be met. These are:

- Inadequate supply of quality planting material,
- Heavy losses caused by several biotic and abiotic stresses, and
- Several unresolved chronic disorders.

As a result, the productivity per unit area is low, resulting in high cost of production. Further, the quality of produce in many cases is far from satisfactory. The post harvest losses continue to be high. Full advantage has yet to be taken of several frontier areas *e.g.*, biotechnology, protected cultivation, computer aided management of inputs, integrated nutrient management, leaf nutrient standards, biofertilizers, integrated pest management and mycorrhiza. There is also need for change both in the content and approach of research which can be taken up in partnership with private sector on aspects like production of hybrids, green house production of flowers, biotechnology, value addition and export.

The future growth of horticulture industry will largely depend on new and globally competitive technologies. As such, ambitious research programme is called for in horticultural crops in the following thrust areas.

Genetic Resources

- Introduce fruits like mangosteen, durian, rambutan, longan, macadamia and berries not yet commercially exploited in India.
- Widen genetic base in mango (*Mangifera* species of South East Asia), Citrus (newly developed rootstocks), apple (scab resistant cultivars), guava (coloured varieties), papaya (species and varieties) and litchi (varieties from South East Asia and China). Also, widen genetic base for improvement of medicinal and aromatic plants required by pharmaceutical industries.
- Enrich germplasm collections with species/cultivars which are high yielding and resistant/tolerant to different biotic and abiotic stresses.
- Exploit full potential for cultivation of avocado, kiwi and olive.
- Standardize long term techniques for cryopreservation of propagating materials and pollen grains.
- *In situ* conservation of endangered genetic wealth.

Crop Improvement

- Develop dwarf rootstocks/scion varieties for high density planting and export in mango, litchi, sapota, citrus, *ber,* coconut, arecanut, oilpalm and cashewnut.
- Induce and exploit useful genetic changes through mutations in commercial cultivars.
- Develop rootstocks and scion varieties in fruits resistant/ tolerant to major biotic and abiotic stresses, *e.g.* malformation in mango, guava rootstocks for wilt, citrus rootstocks against phytophthora, root wilt resistance in coconut, frost and PRV resistance in papaya, weevil in sweet potato, cassava mosaic and stress tolerance in arecanut and spices.
- Develop F_1 hybrids in vegetable crops for yield, nutritional quality and export.
- Development of virus resistant potato varieties having high tuber dry matter and low accumulation of sugars during low temperature storage and varieties with high temperature tolerance.
- Evolving varieties with bold nuts and higher shelling percentage in cashew.
- Domestication of indigenous medicinal plants with bulk demand both in the modern and traditional systems of medicine.

- Studies on the quality control, seed health, packing and storage of seeds of annual crops.

Crop Production

- Standardise rootstocks for all important fruit crops *e.g.* mango, guava, litchi, sapota, *ber* and walnut.
- Develop horticultural crop based cropping systems for different agroecological regions.
- Develop integrated nutrient management system *i.e.* efficient utilization of chemical fertilizers, use of bio-fertilizers and organic material using leaf nutrient standards.
- Standardize water management practices in major crops including micro-irrigation and fertigation.
- Develop techniques of organic farming for export oriented horticultural crops.
- Standardize practices for production of horticulture crops particularly flowers and vegetables under protected cultivation involving computer aided supply of inputs.
- Standardize production technology for quality crops for export *e.g.* mango, grape, litchi, cashew, potato and spices.

Post Harvest Management

- Reduce losses occurring during harvesting, storage and transport and improve shelf life of perishable horticultural commodities.
- Conduct basic and applied research in CA/modified atmospheric storage of high value perishable crops.
- Improve indigenous low cost storage systems developed for onion, potato, fruits and vegetable crops.
- Develop techniques for bulk preservation of fruit pulps, improve drying systems for raisins, mushroom and different vegetables, and standardize technique for frozen vegetables.
- Initiate post harvest research on ornamentals with special emphasis on export market.
- Utilize wastes for development of economically viable products.

Biotechnology

- To standardise *in vitro* culture techniques for mass multiplication of rootstock/scion of difficult to micro propagate plants like mango, guava, sapota, litchi, walnut, date palm, coconut and oil palm.

- Development of protocols for rapid propagation of seedless water melon; vegetable crops with male sterility and self incompatibility barriers *e.g.*, tomato and capsicum and selected ornamental crops.
- Use of anther culture system for production of di-haploids for integration with breeding programmes.
- To identify molecular markers based on RAPD and RFLP in important varieties of horticultural crops.
- Develop transgenic plants with endogenous resistance to insect pests *e.g.* bacterial canker in kagzi lime, salt and drought tolerance in tomato and capsicum.

National Problems

Intensify interdisciplinary research to find solutions to nationally important problems *e.g.*, malformation and irregular bearing in mango, wilt in guava, die back in citrus and root wilt in coconut.

India has a good natural resource base, an adequate R&D infrastructure and excellence in several areas. As a result, the horticultural scenario of the country has been changing fast. Both production and productivity of several crops has increased manifold and India can boast itself as a leading horticultural country of the world. Many new crops have been introduced and many others have adapted to non conventional areas. Some other crops are under adaptive trials. A number of turn key projects in mushrooms and flower production have been established. Near self-sufficiency has been achieved in many crops. Export of fresh as well as processed fruits has been increasing.

The demand of horticulture produce is on the rise due to increasing population, changing food habits, realisation of high nutritional value of horticultural crops and greater emphasis on value addition and export. However, several challenges are yet to be met. These are, fast eroding genepool, fast population build up, shrinking land and other natural resources, serious production constraints, both biotic and abiotic and huge post harvest losses. Further, in the era of globalisation, produce has to be of international quality and globally competitive. The future expansion of horticulture has to be in and semiarid areas and on under utilized crops. While the impact of green revolution in

India was felt mainly in assured irrigation areas, horticultural crop production has brought prosperity even in and semiarid areas. Horticulture is no longer a leisurely avocation and is fast assuming position of a vibrant commercial venture. Nature has placed India in a state of advantage and it is now on us horticulturists to work towards ushering in a GOLDEN REVOLUTION in years to come in India.

Research Preparedness for Accelerated Growth of Horticulture in India

The horticulture scenario of the country is rapidly changing. The production and productivity of horticultural crops have increased manifold. Production of fruits and vegetables has tripled in the last 50 years. The productivity has gone up by three times in banana and by 2.5 times in potato. Today horticultural crops cover about 25 per cent of total agricultural exports of the country. The corporate sector is also showing greater interest in horticulture. A major shift in consumption pattern of fresh and processed fruits and vegetables is expected in the coming century. There will be greater technology adoption both in traditional horticultural enterprise as well as in commercial horticulture sectors. Diversification and value addition will be the key words in the Indian horticulture in the 21st Century.

Horticulture research in India is about four decades old. Systematic research on fruit, vegetable and ornamental crops began in 1954 with the initiation of independent institutions and programmes. The research agenda is designed relevant to national plans and priorities for the horticulture development. Today, eight ICAR institutes with 27 regional stations, 1 project directorate, 10 national research centres, 16 all India coordinated research projects (AICRPS) with 223 research stations, 1 full-fledged university of horticulture, 25 state agricultural universities and 7 multi-disciplinary institutes of the ICAR are engaged in horticulture research. In addition, a few R&D establishments of crop/commodity boards and private sectors are providing research support to Indian horticulture. Research system in horticulture is now geared to provide necessary technological support to the expanding horticultural industry.

The research efforts in the past were mainly concentrated on crop improvement, propagation of seed/planting material, agrotechniques, crop protection and post harvest management. Some of the improved technologies developed are enumerated below.

Technology Generation

Varietal Development

Among the fruit crops, improved high yielding mango varieties, Mallika, Amrapali, Ratna, Sindhu, Arka Aruna, Arka Puneet, Dashehari-51 and hybrids namely CISH-M-1 and CISH-M-2 have been developed. Mallika is coming up in southern states like Karnataka and Amrapali is performing well in eastern India. Dashehari-51, a regular bearing cultivar with about 38 % higher productivity than the normal Dashehari, has been identified after 14 years of rigorous selection. In guava, three selections, namely Lalit, CISH-G-1 and CISH-G-2 have been developed for domestic and export markets.

The fruits of Lalit are of medium size weighing about 150 g each and suitable for both table and processing purposes. In banana, high yielding hybrids like FHIA-01 and FHIA-03 are promising for replacing varieties, Panchananda and Bluggoe, respectively. Cultivar, Saba is found promising under sodic soils. In addition, high yielding varieties like Col, Hl and H2 have been developed. In grape, superior and high yielding varieties have been developed *e.g.* Beauty Seedless and Pusa Seedless and early ripening variety, Perlette for cultivation under North Indian conditions and Anab-e-Shahi, Dilkhush, Thompson Seedless, Tas-A-Ganesh, Sonaka, Bangalore Blue and Pachadraksha for south Indian conditions.

In citrus, high yielding and cluster bearing varieties of acid lime have been developed. *e.g.*, Rough Lemon, Rangpur Lime, Pramaini, Vikram PKM-1. Trifoliate oranges, namely Flying Dragon and Rich 16-6 are dwarfing types. In papaya, high yielding superior varieties both for table purpose and papain production have been developed. *e.g.*, Co-1 to Co-7, Coorg Honey Dew, Pusa Delicious, Pusa Majesty, Pusa Giant and Pusa Nanha. In apple, superior hybrids have been developed. *e.g.*, Lal Ambari, Sunehari. Red Spur, Star Crimson, Golden Spur, Red Chief, Oregon Spur, Skyline Supreme and Vance Delicious have been identified. Tissue culture protocols for micro-propagation of two commercial varieties have been developed.

Among the vegetable crops, more than 130 open pollinated varieties, 36 hybrids, 3 synthetics and 29 resistant varieties of 20 vegetable crops have been developed and released for cultivation in different agro-climatic regions. These include 40 in tomato, 45 in brinjal, 13 in cauliflower, 12 in chillies, 20 in pea, 9 in musk melon, 16 in onion and 44 in other crops.

In potato, 33 high yielding varieties have been developed indigenously for large scale cultivation in different regions. Kufri Ashoka and Kufri Pukhraj mature in 75 days. Kufri Jawahar and Kufri Satluj are field resistant to late blight. Kufri Jawahar has most ideal plant type for inter-cropping. Kufri Chipsona-1 and Kufri Chipsona-2 have been developed with excellent processing attributes, comparable to exotic varieties. Kufri Swarna resistant to golden nematode is ideal for Nilgiri Hills.

In tuber crops, improved varieties of different tuber crops have been recommended/released for commercial cultivation. These includes 9 varieties of cassava, 15 varieties of sweet potato, 6 varieties of colocasia, 3 varieties each of greater yam and lesser yam, 1 each of *Amorphophallus,* taro and yambean. Cassava varieties, Sree Visakham and Sree Prakash are popular in Kerala. Triploid clone, Sree Harsha with high dry matter and starch content is suitable for industrial belt of Tamil Nadu. Two early maturing varieties Sree Jaya and Sree Vijya have been released for culinary purposes. Elephant foot yam variety, Am-15 has been released with high yield potential of 41 t/ ha.

Among the plantation and spice crops, India is the first country to exploit hybrid vigour in coconut. Twelve hybrids involving tall and dwarf parents and 4 varieties have been released for commercial cultivation. These varieties yield 21 to 89 % more than the local cultivars. Some of the released varieties like Chandra Kalpa and Pratap (Banawali Green Round) are receiving wide acceptance of farmers. Chowghat Green Dwarf variety is good for tender coconut purpose. In arecanut, 4 high yielding varieties, namely, Mangala, Sumangala, Sreemangala and Mohitnagar have been developed, giving about 30 % higher yield than the local cultivars.

In oil palm, first efforts for improvement were made by producing Tenera hybrids using Pisifera pollen imported from Nigeria. Dura x Pisifera hybrids are field tested in East and West Godavari districts, Khammam and Krishna districts of Andhra Pradesh, with yield potential of 20-25 tonnes/ha of FFB from the fifth year. In cashew, 22 region specific selections and 12 hybrids with yield potential of 1.5-2 tonnes of raw nuts/ha have been produced and released for commercial cultivation.

The present standards fixed for cashew varieties include export grade kernels of W-210 to W-240 and at least one tonne per ha yield with 30 per cent shelling. In black pepper, 6 varieties, namely,

Sreekara, Subhakara, Palode-2, Panniyur-2, Panniyur-4, Panchami and Pournami, and 2 hybrids *viz.* Panniyur-1 and Panniyur-4 have been developed. In cardamom, a number of improved varieties have been developed and released for commercial cultivation *e.g.,* CCS-1, Mudigere-1, PV-1, ICRI-1 and ICRI-2. In ginger, varieties like Suprabha, Suruchi, Suravi and Varada have been developed. In turmeric, several varieties *viz.*, Co-1, Krishna, Sugandham, BSR-1, Suvarna, Roma, Suroma, Rajendra Sonia, Sugana, Sudarshana, Ranga, Rasmi, Prabha, Prathiba, Mega Turmeric and RCT-1 have been developed with yield potential of up to 44 tonnes of fresh rhizomes per ha. Three high yielding cinnamon lines, namely, Navashree, Nithyashree and Konkan Tej have been released for cultivation.

Seed/planting Material Propagation

In most of the fruit crops, vegetative propagation techniques have been standardized. Soft wood grafting has been standardized for mango, sapota, custard-apple and jackfruit. Other vegetative propagation techniques have been developed for *ber, aonla,* jackfruit, custard-apple and *bael.* In mango, veneer grafting and stone-grafting is practised commercially. Mango variety, Vellaikolumban is suitable semi-dwarfing rootstock for Alphonso. Old unproductive mango trees can be rejuvenated successfully by pruning the 4th order branches during December-January. Flowering and fruiting are regular in pruned trees. For mandarin orange, Rangpur lime is a drought hardy rootstock. In grapes, Dogridge and Salt Creek (Ramsey) are suitable for minimizing adverse effects of soil salinity on Thompson Seedless.

A tissue culture technique for mass multiplication of Dogridge has been standardised. In banana, sword suckers of 700-1000 g are optimum. Rhizomes with active lateral buds and dead central buds are preferred for distant transportation in western India. Double paring and shade drying followed by dipping in Monocrotophos (0.5 %) and Bavistin (0.2 %) is recommended to disinfect nematodes and soil borne fungi. Tissue cultured banana plants are now commercially adopted for their uniformity in flowering and produce. Shoot-tip grafting technique in citrus has been considerably advanced.

In vegetable crops, seed production of over 120 open pollinated high yielding varieties of different vegetables has been well established in the country. Hybrid seed production has become easier with the development of male sterile lines in tomato, self incompatible lines in cauliflower and gynoecious lines in cucumber and muskmelon. In brinjal, functional male sterility controlled by a single recessive gene has been identified.

Temperature barrier in cole crops (cabbage and cauliflower) has been overcome by developing heat tolerant hybrids. It is now possible to cultivate cabbage and cauliflower in southern India. Development of tomato varieties resistant to bacterial wilt has made their cultivation successful in non-traditional areas. Onion seed production technology for cultivation in kharif season has been developed for north Indian states especially, Haryana, Punjab and western Uttar Pradesh. Seed Plot Technique has been developed for production of disease-free potato seed in plains. It is widely adopted by farmers. A new technology for raising commercial crop of potato using 'True Potato Seed' (TPS) has been developed and standardized as supplementary technology to the traditional tuber grown crop. Two TPS populations, TPS-C-3 and HPS-113 are recommended for commercial production in Bihar, Gujarat, Tripura and West Bengal. Micropropagation protocols have been developed in banana, oil palm, cashew, black pepper, ginger, *etc.* Seed gardens of Tall (T) x Dwarf (D) and D x T hybrids have been established for production of coconut hybrids.

Agrotechniques in Horticulture

In fruit crops, improved agrotechniques developed have helped the farmers in improving the productivity and quality of produce. Soil application of paclobutrazol (4 g/tree) increase flowering and fruiting in mango on commercial scale in coastal Maharashtra. It also controls irregular bearing in cultivar Dashehari. Spray of NAA @ 200 ppm in October is recommended for control of malformation. Heavy fruit drop at maturity in cultivar Langra can be controlled by spraying NAA (20 ppm). In guava, double spray of 10 and 20 % urea on cultivars, Allahabad Safeda and Sardar twice at bloom eliminate poor quality rainy season crop and increases winter season yield by 3 and 4 times, respectively. Application of neem coated urea (800 g/plants) yields 98 kg fruits/plant in variety, Sardar compared to 37 kg from untreated plants. In banana, high density planting (4550 plants/ha) yield up to 174 t/ha.

Adoption of improved technology in Maharashtra has resulted in fruit yield increase up to 52 t/ha. In citrus, two grafting methods using inverted 'T' cut and apical triangle cut have been developed with overall success of around 36% using either method. For accelerating the survival of growth, shoot tip grafts, the successful grafts are double grafted (side grafted) on vigorous greenhouse grown

Rough Lemon and Rangpur Lime seedlings. Rangpur Lime rootstock is found superior for sweet oranges and mandarins. In papaya, closer spacing of 1.4 x 1.4 m is recommended for high yield. Drip irrigation techniques have been standardized. In banana, it has resulted in production gain (60-70%) and early harvesting (40-50 days), besides improved water efficiency. Likewise, in grapes higher yields have been obtained with better water use efficiency (11 %).

In vegetable crops, improved production technology has been developed for major crops. Drip irrigation is economical in tomato and brinjal. In cucumber, replenishment of evaporation loss through irrigation resulted in maximization of yield of quality fruits. Drip irrigation in watermelon provided 33% higher yield with a water saving of 40%. Nutrient requirements and fertiliser schedules have been worked out crop-wise and recommended for different agro-climatic regions. In leguminous vegetables, high N depresses nodulation. The VAM fungi increases P availability to plants.

In all leguminous vegetables, inoculation of the VAM fungi along with *Rhizobium* culture is beneficial. Production technologies for *kharif* onion in northern India and long day type onions for high altitudes have been standardized. Pendimethalin (Stomp) has been found effective in controlling weeds in tomato, brinjal, chilli, bell pepper and okra. In potato, a number of potato-based multiple and inter cropping systems have been developed for different potato growing regions. Intercrop combinations with sugarcane in Maharashtra, wheat in Chhota Nagpur area and linseed in central Uttar Pradesh are found remunerative. A suitable method of urea application has also been worked out.

In tuber crops, short duration legumes *viz.*, groundnut and French bean and cowpea can be successfully inter-cropped with cassava. Short-duration cassava, Sree Prakash is ideal in double cropped rice fields. Studies on cassava-based multiple cropping systems involving banana, coconut, *Leucaena* and Eucalyptus, have shown banana-cassava combination to give maximum root yield. Banana and coconut combination reduces soil loss and surface run-off considerably. *Dioscorea* and elephant foot yam with banana, Nendran can generate an additional income of Rs 20,000/- over the sole crop of banana. For Inter-cropping *Dioscorea* in coconut garden, the ideal planting density is 9000 plants/ha. When *Amorphophallus* is raised as an inter-crop in coconut garden, one third dose of recommended fertilizers is sufficient. Inoculation with VAM fungi in cassava give about 15-20% increase in yield.

Among the plantation and spice crops, density of 175 coconut palms/ha (7.5 m x 7.5 m spacing) is found ideal. In general, NPK application of 500:320:1200 g/palm/year is found optimum. A multi-storied cropping system involving black pepper trained on coconut trees, and cocoa in between the rows of coconut and pineapple in the ground floor has been found ideal for exploiting light, soil and air spaces. In arecanut, application of NPK (100:40:140 g) and green leaf (14 kg) per palm per year is recommended for coastal regions of Kerala and Karnataka and for plains of West Bengal, Karnataka and Assam.

In oil palm, application of NPK (1200:600:1200 g/palm) is found to give 17.1 tonne of FFB per ha. In black pepper, rapid methods for production of rooted cuttings of pepper have been developed and a commercial protocol has been standardised for micropropagation of black pepper. Ginger yield could be increased up to 33 percent by application of neem cake at the rate of two tonnes per ha and the fertiliser schedule of 75 kg each of N, P_2O_5 and K_2O. Technology for storage of ginger seed rhizome is standardised and recommended.

Protection Technologies

For major fruit crops, plant protection schedules have been developed for the control of significant insect-pests for wider adoption. In the recent years, research efforts are directed to devise eco-friendly, economical and long lasting control measures. Success has been achieved in biological control of mealy bugs in mango and guava. The *Beauveria bassiana* has been found killing mango mealy bug and hopper. In grapes, integrated management of *Spodoptera* caterpillar involving light and pheromone traps, NPV and neem based insecticides and biological control of mealy bug by the beetle *Cryptolaemus montrouzieri* have been standardized. Studies on pesticide residues have resulted in working out of safe-waiting periods for harvesting and consumption of fruits.

In vegetable crops, about 50 improved measures for efficient management of diseases and 23 for insect-pests have been worked out and popularized in different agro-climatic regions in the country. Integrated pest management (IPM) for controlling diamond back moth on cabbage through a trap crop like mustard has been demonstrated. Fruit borer *(H. armigera)* on tomato can be controlled by the release of *Trichogramma pretiosum* alone and in combination with HaNPV. In potato, integrated management schedules for control of bacterial wilt and tuber moth have been developed. A late blight forecasting system has been developed for the hills.

Among the tuber crops, the major diseases affecting tuber crops are cassava mosaic and brown leaf spot in cassava, *Phytophthora* leaf blight in colocasia, *Fusarium* wilt in elephant yam and virus diseases of sweet potato.

Foliar sprays of Bavistin (0.1%) combined with disodium and dipotassium phosphates (100 ppm) and calcium sulphate at 15-day interval was found to check foliar diseases in sweet potato. The major pests include spider mites, scale insects and white fly on cassava, weevil on sweet potato, defoliators, aphids and mites on colocasia, and scales and mealy bugs on yams and elephant yam. Cultural methods for weevil control include clean cultivation, destruction of alternate hosts and timely harvest. An effective IPM package using synthetic sex pheromone has been developed. Control measures involving insecticides have been evolved for the control of pests of other tuber crops.

Among the plantation and spice crops, bud rot of coconut caused by *Phytophthora palmivora* can be effectively controlled by spraying Bordeaux mixture. Calyxin root feeding and drenching of soil with 1% Bordeaux mixture along with neem cake application @ 5 kg per palm per year is recommended for controlling Thanjavur wilt disease reported in Tamil Nadu, Andhra Pradesh and Karnataka.

A package of practices has been developed for managing mycoplasma like organisms (MLOs) in root wilt affected coconut palms in Kerala and Thatipaka disease affected palms in Andhra Pradesh.

Eradication of all root wilt affected palms is recommended. In cashew, tea mosquito bug (TMB) can be effectively controlled through a schedule of spray coinciding with flushing, flowering and fruiting. For effective control of stem and root borer infestation, constant monitoring and adoption of strict sanitation in the plantations coupled with prophylactic application of coal tar and kerosene in the ratio of 1:2 on trunks are recommended. In black pepper, spraying Bordeaux mixture (1 %) and drenching the soil with copper oxychloride (0.2 %) is found effective in managing *Phytophthora* foot rot.

Post Harvest Management Technologies

The post harvest handling of fruits and vegetables accounts for 20-30% of losses at different stages of storage, grading, packing, transport and finally at marketing as a fresh produce or in processed form. A number of improved technologies have been developed for commercial exploitation. An on-farm, low cost, environment friendly cool chamber, Zero Energy Cool Chamber has been developed using locally available material.

The principle of evaporative cooling reduces the inside temperature by as much as 17-18 °C and keeps the relative humidity above 90% during peak summer. The chamber increases the shelf life and reduces PLW of banana, mango, orange lime, grape fruit, tomato and potato in different situations in India. Maturity standards for mango, guava, grape, litchi and *ber* and chemical treatments for regulation of ripening in mango, sapota and banana have been standardised.

Optimum storage temperatures worked out for several fruits, vegetables and tuber crops. A mango harvester, fruit peeler, hand and pedal operator cassava chipping machines, harvesting tools (5-14 times efficient), coconut dehusking machine, implements for mechanization of potato cultivation and other crops have been developed. A number of improved technologies have been developed for commercial exploitation *viz.*, tent type foldable solar dryer, packaging boxes for distant transportation of apple, mango, citrus and plum, production of value added products-pectin from peel and flour from mango fruit kernel, production of fruit post carbonated beverages *etc.*

Production Constraints

In spite of great strides made, the productivity of horticultural crops, in general, is still quite low and the post harvest losses particularly of perishable commodities, are considerable. Improvement in quality standards of the produce and their marketing are essential to increase our share in the global market. The research agenda in horticulture is by design relevant to national plans and priorities and research programmes are normally formulated keeping in view the thrust areas in development. The major technology related constraints contributing to low productivity of horticultural crops and inferior quality of produce are:

- Vast majority of holdings are small and un-irrigated.
- Large tracts of low and unproductive plantations needing replacement/rejuvenation.
- Low productivity of crops due to inferior genetic stocks and poor management.
- Inadequate supply of quality planting materials of improved varieties.

- High incidence of pests and diseases.
- Heavy post harvest losses and low utilization in processing sector.

For addressing the above constraints, research institutions are engaged in both basic and applied research. While formulating research strategies some of the inherent weaknesses associated with perennial tree crops and certain perpetual problems in Indian horticulture must be kept in mind. They are:

- Long period required for development of improved genotypes. Application of biotechnological tools/methods in horticultural crops is still in its early stage of development in the country.
- Chronic production problems due to major disorders like alternate bearing, malformation and spongy tissue in mango, guava wilt, citrus decline, root wilt in coconut, viral disease in vegetables, *Phytophthora* diseases in large number of crops *etc.* remained largely unresolved.
- Lack of advanced technologies for post harvest handling, processing and marketing of produce.

Losses caused by biotic stresses are very high and due to pesticide residue problems development of eco-friendly IPM strategy is more relevant in horticulture. There is a threat for loss of valuable genetic resources, if measures are not taken for their conservation.

Wastelands and hilly terrains being the potential future expansion areas, matching technologies for dry land and hill horticulture need to be developed. Counter seasonal advantages from diverse agro-climatic situations provide strength for extended availability of horticultural crops round the year and such potentials can be harnessed only with relevant research support.

4

Methods of Weed Control

Weed management decisions vary according to plant life cycles, infestation size, environmental parameters and management objectives. Hand-pulling a perennial weed species such as Canada thistle or leafy spurge is a futile effort, but very effective for control of a biennial plant such as diffuse knapweed. Releasing biocontrol insects for control of dalmatian toadflax on a 500 acre property is wise, but relying on insects for controlling small patches on a 40 acre pasture is inefficient. Successful weed management requires proper plant identification, selection of effective management methods and monitoring the effects over time.

Preventive Control

Prevention is the most essential aspect of weed management. Once a noxious weed infestation becomes established, any increase in size and density creates increasingly more expensive management efforts.

Awareness of weed seed sources and plant identification is a must. Feel free to call the Weed District office for help with identification or to set up a site visit, and recognize:

- Weed seed can be spread from neighbouring properties, adjacent road rights-of-way and trails. Direct sources are often livestock, manure, seed, hay, vehicles and equipment.
- Disturbed ground is most vulnerable to weed invasion; new roads, pipelines and other sites where competitive vegetation has been removed. With no restoration weeds will likely appear.
- Early detection and rapid response saves time and money. Aggressive management action on small, newly established

infestations can result in eradication. "An ounce of prevention is worth a pound of cure".

Cultural Control

Cultural control, the establishment of competitive and desired vegetation, prevents or slows down invasion by weedy species and is a key component of successful weed management. Weeds are typically opportunistic and readily invade disturbed sites. Impacts from road construction, intensive livestock grazing, densely populated prairie dog colonies and other disturbances that damage or remove desirable and competitive vegetation create sites for noxious weed invasion.

Controlling weeds on such sites can be futile without vegetative restoration, as weeds will readily re-invade the disturbed area. Establishment of grassland or pasture can be challenging. Success often depends on proper species selection suitable for a particular soil type, moisture regime and growing season. Other factors such as soil compaction, seeding depth, time of year, and weed control during establishment can be critical to success. Whether seeding to native plants or pasture grasses, it's important to consult with a commercial seeder, CSU Extension or other experts prior to investing time and money.

Chemical Control

Herbicide application can provide the most effective and time-efficient method of managing weeds. Numerous herbicides are available that provide effective weed control and are selective in that grasses are not injured. Along with herbicide use is user responsibility and compliance with all product label requirements for herbicide handling, use, and cleanup. Always read the label and keep in mind the label is legally binding. When using herbicides be mindful of proximity to water, trees, shrubs and other desirable vegetation. The Herbicide Reference Guide lists products commonly used for control of noxious weeds on small acreage properties.

The guide provides information necessary to help decide which herbicide is most appropriate for a particular weed in a particular setting, but is not the complete label. Herbicides are applied by spot spraying-single nozzle application targeting individual plants, or broadcast spraying-multiple nozzles covering an entire area.

Whatever method is used, calibration of spray equipment (gallons per acre spray output) is essential for accurate delivery and mixing calculations. Estimating or guessing sprayer output can lead to misapplication which either injures non-target plants or results in failure to control the target weed species.

Biological Control

The goal of biological control is not eradication, but the use of living agents to suppress vigour and spread of weeds. Such agents can be insects, bacteria, fungi, or grazing animals such as sheep, goats, cattle or horses. Grazing produces results similar to mowing, and bacteria and fungi are seldom available for noxious weed management. Biological control is most commonly thought of as 'insect biocontrol'.

USDA-APHIS (United States Department of Agriculture-Animal and Plant Health Inspection Service), is the federal agency responsible for authorizing the screening and importation of biocontrol insects. APHIS conducts intensive multi-year screening programs assuring an insect agent's host specificity (feeds only on target weed species, not other plants) prior to approval for release.

Biological weed control through insect/plant interactions is an important component of the County's weed management program. Insect agents, proven to be effective, are utilized in cases where eradication is impractical due to the vastness or inaccessibility of an infestation, and where other methods of management are not feasible.

Insect agents typically require 3-5 years for establishment and can limit the spread and density of target weed species by feeding on leaves, stems, roots and/or seed heads. One must realize that eradication of a weed cannot be attained through insect biocontrol.

The most effective scenario is a weed infestation reduced to a 'tolerable level', a level where the insect agents are significantly limiting distribution and abundance of the target weed species and the weed density is no longer considered detrimental to the desired plant community. Some biocontrol insects proven to be successful in Larimer County are:

- Bindweed mites-field bindweed
- Flea beetles-leafy spurge
- Mecinus janthinus-dalmatian toadflax
- Larinus minutus and Cyphocleonus achates-diffuse knapweed.

Mechanical Control

Mechanical control consists of methods that kill or suppress weeds through physical disruption. Such methods include pulling, digging, disking, plowing and mowing. Success of various mechanical control methods is dependant on the life cycle of the target weed species.

- Hand pulling and digging are effective on annual and biennial species such as kochia, musk thistle, and diffuse knapweed. It is important to remove the upper 2-3 inches of taproot to prevent re-growth. Hand pulling or digging a perennial weed such as leafy spurge can be a futile effort unless one has the time necessary to diligently dig or pull re-growth over several seasons.
- Shallow tillage with a disk or sweep is effective for controlling annual species such as cheatgrass or kochia, but can actually be counter productive if trying to control perennial weeds such as Canada thistle, field bindweed, leafy spurge or Russian knapweed. Perennial root systems often have meristematic buds that can set roots and produce a new plant from root segments deposited on the soil surface. Shallow tillage of perennial weeds can result in a larger, denser and more uniform infestation than the initial patch.
- Moldboard plowing (complete turnover of the top 10-12 inches of soil) disrupts underground root systems and buries seed from the surface to a depth too deep to germinate. This type of tillage is seldom feasible to practice on a regular basis.
- Mowing is a suppression measure that can prevent or decrease seed head production. Mowed weeds will re-grow and set seed from a reduced height so a combined control method is necessary to be effective. Mowing causes perennial plants to weaken when forced to send up carbohydrates from underground root reserves to nourish re-growth. So mowing a perennial weed such as Canada thistle a couple of times during the summer can significantly weaken the plants, and when combined with a fall herbicide application, provides excellent control.

Control of Weed

Weeds are the plants, which grow where they are not wanted. They grow in the fields where they compete with crops for water, soil nutrients, light and space and thus reduce crop yields. They also harbour insects pests and microorganism. Certain weeds release into the soil the inhibitors or poisonous substances which are harmful to the plants, human beings and live-stocks.

They increase the expenditure on labour and equipment, render harvesting difficult, and reduce the quality and marketability of agricultural produce. They block the drainage and impede the flow of water in canals and water-transport channels and their growth in the rivers renders navigation very difficult. The dense growth of weeds in water pollutes the water because they deoxygenate the water and kill the fish.

Natural Powers of Weeds

Weed seeds germinate earlier; their seedlings grow faster; they flower earlier; and form seeds in profusion; and mature ahead of the crop they infest. Nature has bestowed these qualities on weeds so that there seeds are collected unwarily along with the produce of the crop at harvest and get distributed to other places where the produce may be taken.

They have the remarkable capacity to germinate under varied conditions, but very characteristically they are season-bound and the peak period of germination always takes place in certain seasons in regular succession year after year. Another charecteristic of the weed seeds is the possession by them of the phenomenon of dormancy which is an intrensic physiological power of the seed to resist germination even under favourable conditions, and also the seeds do not lose their viability for years even under adverse conditions.

Types of Weeds

Weeds belong to the class Angiospermae (flowering plants) which have two subclasses: Monocotyledoneae (monocots) and Dicotyledoneae (dicots). On the basis of the habitats, they are divided into terrestrial and aquatic categories, and on the basis of the duration of life they are divided into annuals, biennials and perenials. Annuals live and produce their seeds in a single growing season. Biennials need two growing seasons; in one season they pass through their vegetative or rosette stage, followed by reproductive stage in the next season. The multiplication of both the annuals and the biennials is through seed. Perenials live indefinitely and are propagated not only through seeds but often vegetatively through underground structures, such as rhizomes, stolons, bulbs and tubers.

Perenials are of two types: the simple and the creeping. The former multiply only through seeds. They have no normal means of spreading vegetatively. However if they are injured or cut, the severed portion produce new plants. Creeping perenials are spread by creeping roots, creeping above-ground stems (stolons), and creeping underground stems (rhizomes). Some weeds propogate themselves by means of tubers which are modified rhizomes adapted for the storage of food.

Aquatic weeds (hydrophytes or water inhabitants) are classified into three types, viz. submerged, emersed and floating. Submerged aquatics are anchored to the bottom of the habitats, e.g. a ditch, and grow entirely beneath the surface of the water. Emersed once have their roots beneath the surface of the water, but the leaves and stems are above the water-line. Floating weeds or surfaced aquatics either float freely on the water or float only in a limited area.

There is still another group of weed known as the parasitic weeds. These are of two kinds: the total parasite and the partial parasites. These weeds parasitize certain host plants, which they directly attack and deprive them of water, nutrients and assimilates.

Principles of Crop Production and Weed Destruction

Crop-weed competition. In non-irrigated areas, the competition between weeds and crops is largely for water. The transpiration coeffiicients for *kunda (Ischaemumpilosum), hariali(Cynodon dactylon)* and *Tephrosia pupurea* are 556, 813 and 1,108 respectively, whereas this coefficient is only 430 for *jawar (Sorghum vulgare).* A saving of 750 to 1,250 tonnes of water per hectare of soil, forming a one-metre deep column, is possible by keeping the soil free from weeds.

In irrigated tracts, the competition is severe for nutrients. The unchecked growth of weeds, in a wheat fields measuring one hectare, removes about 20 kg of nitrogen, reducing the grain yield by about twelve quintals. The mineral requirements of weeds are high. Wild mustard *(Brassicasinensis)* requires twice as much nitrogen, twice as much phosphorus, four times as much potash and four times as much water as a cultivated oat plant. Weeds in a fallow land deplete the soils of both moisture and nutrients. *Pohli (Carthamus oxyacantha)* removes about 60 kg of nitrogen from one hectare.

Weeds generate a formidable competition with crop plants for light. One of the secrets behind the high-yield potential of the dwarf varieties of crop plants is that they can capture solar energy more efficiently. Unchecked weed growth in the fields cropped with dwarf varieties will intercept solar radiation to the detriment of the crop plants. Likewise, the competition for space that the weed offer to the valuable crop plants may assume, at times, so alarming proportion as to smother the latter.

The identification of weeds. Both the nature of the weeds and that of the crops which they infest, influence the action of the weedicides. For example, the same types of weeds occur in maize and cotton which are *kharif* crops. 2,4-D can be used for killing the weeds in maize, but this weedicide should never be used on cotton which is highly sensitive to it. The successful application of chemicals to smother the weeds in growing crops is determined by the knowledge one has regarding the biology of both the crops and the weeds. The degree of success from the adoption of physical methods of weed removal is determined largely by the knowledge one has of the peak periods when the seeds of weeds are formed and of the nature or duration of their dormancy.

Preventive measures: Nature has provided weeds with a number of devices that help them to be disseminated widely.The agencies that facilitate the dispersal of weed seeds far and wide are water, wind and animals, including man. The troubles that weeds create in crops, soil and water are summed up in the adage "one year of seedling is seven years of weeding". To avoid such a situation, a wise step is to follow the principle "prevention is better than cure".

As weed seeds are so readily dispersed by natural agencies and by the farmer himself, it is important to prevent weeds, whether in crops, on borders or bunds, in fences or in irrigation-channels, from flowering and setting seed. Preventive methods consist in sowing crop seeds not contaminated with weed seeds,using manure and irrigation water not laden with them and the enforcement of weed control laws and seed-certification measures.

Control methods. Measures against weeds comprise mechanical(cultivation and moving), cultural or cropping, biological and chemical means. Each of these methods has certain merits and a prudent farmer can make use of one means or a combination of them to control weeds efficiently and economically.

Mechanical methods. Hand-pulling or hand-weeding, hoeing, tilling, mowing, burning, flooding, smothering, etc. are examples of physical methods of weed control, involving the use of physical energy through implements,either manual, bullock-drawn or power-operated. Hand-weeding is the most efficient method,but it is back-breaking,time-consuming and costly. With the gradual industrialization of our country, coupled with the raising of standards of living and literacy, manual labour is becoming scarce. Further, the high wages paid to the hired labour narrow down the profits of the cultivator.

Tillage is a practical and economical method of controlling annual weeds. The plough (disc,mouldboard,etc.), the harrow (disc,spike and spring-toothed) and the cultivator (duck-foot, blade,*bakhar*, etc.) are implements in use to eradicate weeds.

Cultural or cropping methods. Weeds under many conditions are better compititors than crop plants for light, water, nutrients and soil space. However, farming practices are capable of changing the condition in such a way as to enable the crop plants to compete with weeds successfully or to reduce their interference to the minimum and thus preventing them from acting as impediments to increased crop production. Seeds with good germination will give the crop a vigorous and close stand and thus enable it to steal a march on the weeds. Varieties which are well adapted to a region will obviously compete better with the weeds than varieties poorly adapted to it. The plant breeder has to evolve quick-growing and short-duration varieties of crop plants with a large leaf area and good branching and the agronomist has to work out the proper seed-rate,depth,time and method of sowing, applying the most effective methods of irrigation and fertilizers and adopt a proper system of rotation.

Some crops can compete better with weeds than others. For instance crops like Sudan grass, sorghum and cowpea are good competitors while crops,such as linseed,groundnut and lentil are poor competitors. The raising of highly competitive crops (also known as smother crops) is useful in reducing weed infestations. One disadvantage from which many smother crops suffer is that their seedlings grow slowly. If these crops are grown for seed, some measures have necessarily to be taken to control weeds in the early growth of these crops. However, if they are meant for fodder, the thick sowing resulting in dense growth smothers the weeds.

Chemical methods The controlling of weeds in the growing crops with weedicides increases their yields and ensures the efficient use of irrigation, fertilizers and plant-protection measures, such as the spraying of insecticides and fungicides. The removal of weeds from the growing crops facilitates easy harvesting and gives a high-quality produce without admixture with weed seeds. Chemical weed control can be adopted quite in time and in situations and under conditions which make manual or mechanical weeding difficult. A great advantage of this method lies in killing weeds in the crop row or in the immediate vicinity of crop plants. The chemical method is easier, less time-consuming and less costly than weeding by hired labourers.

A herbicide is any chemical that kills the plants or inhibits their growth. Selective herbicides remove certain weeds from certain crops. The selectivity is not absolute, but is governed by the amount of the chemical applied, the way it is applied, the degree of wetting of the foliage, the amount of rainfall following the application, the tolerance of different plants to a specific chemical and the differences in the growth habits of the crops and the weeds. Non-selective herbicides remove a wide range of vegetarian, although plants differ in their susceptibility to any particular chemical.

Selective Herbicides

A. *Foliage Applications.* These applications are made to the leaves of growing plants, usually as sprays, but in a few cases as dust applications.

(i) *Contact herbicides.* Contact herbicides kill only the plant or the portions of the plant that actually come into contact with the chemical. The herbicide to be effective must cover the foliage. Selectivity is dependent upon the differential wetting, differences in forms of the plants, and also upon the placement of the spray. Some examples are dicryl, potassium cyanate, selective weed-oil (carrot oil), sodium arsenite, Solan, Propanil and sulphuric acid.

(ii) *Translocated herbicides.* Translocated herbicides move within the plant,a property that makes them effective in destroying the roots of perennial weeds. A low-volume application is possible in their case. The physiological differences among plants determine the selectivity. Systemic herbicide is also a term used to name a herbicide which is translocated. Examples of this group of herbicides are Caebyne, 2, 4-DB, MCPB, 2, 4-5-T, 2, 4, 5-TP and MSMA.

B. *Soil Applications.* Soil fumigants and soil sterilants fall into this group. They are used where it is desired to remove all plant growth or to keep areas free from plant growth.

(i) *Soil fumigants* These materials are employed for killing all plant growth before sowing more desirable species. They function as a vapour or as a gas that diffuses through the soil and have a relatively short life in the soil as a result of which re-planting in the treated area is possible within four weeks or less. Carbon disulphide, Chloropicrin, Vapam, methylbromide and cyanamide are some soil fumigants in use.

(ii) *Soil sterilants.* These chemicals make the soil sterile (incapacitate it to support vegetation) for varying length of time,depending on the chemical used, the amount applied, rainfall, soil type, organic matter, etc. The inorganic chemicals, borates and chlorates, and the organic compounds, Trizines (Simazine and Atrazine), urea herbicides (Diuron, Monuron, Fenuron, Linuron, Neburon), Uracils (Bromacil, Terbacil) are some soil sterilants.

C. *Aquatic Applications.* A number of chemicals are used for controlling some submerged aquatic weeds by dissolving or emulsifying them in water in canals,ditches,ponds and lakes. Some chemicals in use are aqualin, aromatic solvents, chlorinated benzenes, copper sulphate, Endothal, Fenac, Sodium arsenite and 2,4-D.

Formulation of herbicides. Formulation refers to the way in which in which the basic weed-killing chemicals are prepared for practical use. Herbicides are formulated to be applied as solutions in water or oil, emulsions, wettable powders, granules and dusts.

Solutions of water or oils. The salts of most herbicides are soluble in water. They are dissolved in convenient amounts of water and then sprayed. A few examples are sodium and amine salts of 2, 4-d, 2, 4,5-T, MCPA, amine salts of DNBP, sodium salts of Pentachlorophenol, TCA and Dalapon. The "parent acid" formulations of some of these are soluble in oil (DNBP,PCP). They are often used to increase the toxicity of oil sprays or to fortify the oil.

Emulsions. An emulsion is one liquid dispersed in another liquid, each maintaining its original identity. The two liquids are prevented from reacting with each other by the addition of an emulsifying agent. Ester formulations of 2,4-D are oil-like and form emulsions. Emulsions are milky.

Wettable powder. It is a type of formulation in which a herbicide is absorbed generally on an inert carrier, together with an added surface-acting agent, and finely ground so that it will form a suspension when agitated with water. Simazine, Atrazine, Monuron, Diuron and Neburon are wettable powders.

Granular herbicides. In these formulations the herbicide is absorbed on, mixed with, or impregnated into, a generally inert carrier in such a way that the final product consists of granular particles. Many carriers are used, e.g. clay,sand vermiculite and finely ground plant parts (ground corn cobs, tobacco leaf trash,etc.). Granular materials can be spread by hand or with mechanical spreaders. These materials have advantages over sprays,because water is not needed for application, costlier spraying equipment is dispensed with, and the granules fall off the leaves of valuable plants without causing injury. Another advantage is that as the active weed-killing principle in these formulations is gradually released, they suppress the growth of weeds for long.

Dusts. Insecticides and fungicides are very often made in the form of dusts. However, only a few herbicides are applied as dusts because of the drift hazard.

Calculation of the dosage levels. The quantity of the toxic ingredients, to which the weed-killing property of a chemical is due, is different in different formulations of herbicides. The active ingredient is that part of a chemical formulation which is directly responsible for the herbicidal effects. In some herbicides, the entire molecule may be the active unit. If the chemical is 99 per cent pure, it is considered to have 99 per cent active ingredient. The strength of the wettable powders is expressed in terms of the active ingredient (a.i.). In others, the herbicidal activity is calculated on an acid-equivalent basis. The acid equivalent (a.e.) refers to that part of a formulation which can be theoretically attributed to the acid. In these cases, the acid equivalent is given as the active ingredient. Examples are 2, 4-D, MCPA, MCPB, 2,4-DB, CMPP and Picloram.

A formula for calculating the dosage is given below:

Pounds of product = *dose of the active ingredient required for one acre* 100/ percentage content of active ingredient in the product*

The acid equivalent is given on the label. If the product has 80 per cent a.e., then to apply 2 kg 2*100 of the active ingredients per hectare the quantity of the formulation required is = 2*100/80 = 2.5 kgs 80

Spray Equipment and its Care

Spraying is the commonest method of applying herbicides. Sprayers are classified as low-volume and high-volume sprayers. Low-volume sprayers usually apply less than 400 litres of the total spray per hectare. With aircraft it is possible to apply as little as 80 litres of spray per hectare. Our small agricultural holdings, coupled with the practice of growing mixed crops present difficulties in the spraying of herbicides with aircraft. However, the delivering of about 800 to 1,000 litres of spray per hectare is good because the drift hazard to sensitive crops is reduced to the minimum. Granular formulations of herbicides appear to hold out much promise in eliminating drift hazard.

A special type of small portable sprayer is the logarithmic portable sprayer being used in experiments in the U.S.A. The advantages of field-testing of herbicides with this sprayer are many. It delivers the chemical at a logarithmically decreasing rate, i.e. if a start is made with a 1% solution, after a certain distance, it will 0.5%, again after the same distance it will be 0.25% and so on, until a dilution is reached that is no longer effective. On a single species one can readily ascertain the minimum lethal dose of a chemical, the maximum tolerated dose, and the maximum dose that produces no symptoms. On mixed vegetation, as in a pasture one can read off the minimum lethal dose for each species and hence determine the selectivity of the chemicals.

Many of the herbicides are corrosive. Wettable powders are frequently abrasive, shorten the life of all parts of the equipment, especially the gear pumps, valves, booms, nozzles, etc. Thorough cleaning and rinsing are necessary. It is a rule to clean any sprayer before and immediately after use. Multi-purpose sprayers (the same sprayer used for herbicides, insecticides, fungicides, etc.) must be subjected to very rigid sanitation measures. The equipment should be thoroughly cleaned with water or preferably with 1% ammonia water before it is used for other purposes. When an ester is used, the equipment should be rinsed with kerosene before it is washed with water. After use, fill the sprayer tank and hoes with water overnight and the next day give thorough rinsing with water.

Working with Selective Weed Killers

Some examples are given below for illustrating the broad principles involved in the selectivity of herbicides.

2, 4-D *For Killing Broad-leaved Weeds In Monocot* crops. Grasses have narrow and erect leaves; the leaf surface is rough and waxy; and the growing-points are located in the crown of the plant below the soil level and are protected by the surrounding leaves. Broad-leaved plants have leaves that are wider and grow horizontally; the leaf surfaces are smooth and even; and the growing-points are exposed at the tips of the shoots and in the leaf axils. The droplets of the 2,4-D spray bounce off the leaves of grass and wet them only in spots. On the other hand, the droplets stick to the broad leaves and completely wet them, on account of which 2,4-D becomes toxic to broad-leaved plants.

2,4, 5-T is lethal to wild cucumber, whereas 2,4-D is not. Tracer research directed towards the study of penetration, translocation and metabolism of these two chemicals has indicated that 2,4 5-T remains unaltered in the toxic form inside the plant, thus proving harmful to wild cucumber. 2, 4-DB IN LEGUMES. 2, 4-DB and MCPB are employed for controlling broad-leaved weeds in legumes. These chemicals become toxic only after they are converted into the corresponding 2, 4-D and MCPA. This conversion is brought about by certain enzyme systems and as certain legumes lack them, the weeds and not the legume succumb to these herbicides.

Simazine in Maize. This chemical, as such, is harmful to all plants, but it is used as a herbicide in the case of maize, as this crop possesses a mechanism which converts the chemical into nontoxic hydroxy-simazine.

Linuron In Carrot. A primary reason for the selectivity and lethality of Linuron to carrot and weeds, respectively is due to the differences in absorption, metabolism and phytotoxicity of several of the metabolic derivatives that are formed inside the tissues.

Barban in Wheat and Barley. The selective control of wild oats in wheat and barley with Barban marks a landmark in the development of selective herbicides. What is remarkable is that although wheat, oats and wild oats belong to the category of grasses, Barban is harmful only to wild oats.

Dalapon for Killing Grasses. Dalapon, as a systematic killer of annual and perennial grasses, is a systematic killer of annual and perennial grasses, is a new tool in the hands of the farmers. This herbicide has proved highly resistent to breakdown in plants. Studies have shown that it is most effective on young plants, in active growth, and on plants receiving an adequate water-supply. These charecteristics of Dalapon, added to the nature of its formative effects (tubular leaf formation, excessive tillering, stunting) and the fact that it is more effective in tn the acid form than in the form of a sodium salt, indicate that this herbicide is a plant harmone which acts on grasses in much the same way as 2, 4-D does on broad-leaved plants.

The inability of grass roots to produce buds to regenerate themselves, coupled with their strong tendency to retranslocate this lethal harmone to meristematic regions, including shoot tips and buds, may furnish a clue to the selectivity of this compound. Its tendency not to accumulate in the roots may explain the survival of broad-leaved plants whose roots can produce buds; its ability to continue to move into new buds as soon as they formed may explain its strong phytocidal action on perennial grasses that regenerate themselves solely by buds.

The Selectivity and Lethality of Herbicides

The selectivity of herbicides is largely due to differences in the mechanisms of absorption, translocation or biotransformation between weeds and the resistent crop plants.

Herbicide residues.Herbicides insecticides and fungicides are grouped under the term pesticides. The question of pesticide residue in the final product, that is consumed by man and animals, has been receiving much attention in America where the pest-control measures are employed are employed on an extensive scale. Chemicals are screened through rigorous procedures and the determinations of LD 50 values (lethal dose) are made. If the instructions given on the containers of the products are faithfully followed, there is no possibility of the pesticides leaving toxic residues in food products.

The question of toxic accumulation in food products may be viewed with less alarm in the case of herbicides than in the case of insecticides and fungicides because of the following basic differences in their use. Insecticides and fungicides are dusted or sprayed on the crop plant and not on the weeds; herbicides are sprayed mainly on the weeds and not on the crop plants. Insecticidal, fungicidal sprays have to be repeated many times during the growing season of the crop, whereas only one spray with the herbicide is sufficient to control weeds. In the case of pre-emergence, or pre-planting application of the herbicides, the time-lag between the treatment and the harvesting of the final product renders the toxic chemical innocuous. In the soil, the chemical is detoxified by microorganisms, cultural conditions and some other natural agencies. Inside the crop plant also, the chemical does not remain in the toxic form in which it is applied; it gets broken down to nontoxic forms.

It is necessary to be unduly alarmed while using imported herbicides in our country, because all of them have been tested and found safe for use on crops.

There are ways of minimising the problem of the toxic build-up of herbicide residues in plants and soil. The use of the proper herbicides at the correct dosage level and in the appropriate manner and at the appropriate time essential for avoiding any possible build-up of toxic residues in the soil. The use of low doses of herbicide is one of the surest ways of minimising herbicide residues, as low doses are adequate enough to suppress the weeds during the first 4-8 weeks after sowing the crops. During that period, herbicide would lose all its toxicity, thus leaving no residues to affect the crops raised in the herbicide-treated field, and further the chances of accumulation of toxic quantities of herbicides in the plant tissues or contaminating the economic products prepared from the crops grown in the treated fields, are reduced when low doses of herbicides are employed. Also, fertilizing the crops adequately and taking timely measures against insects and pests would go a long way in making the crop plants resist the toxic action of herbicides. Other steps that are useful in minimizing the residue problems consists in not disturbing the soil for 4-6 weeks after herbicide treatment; giving a deep ploughing after harvesting a herbicide-treated crop; adding large quantities of organic matter, such as farmyard manure to the soil and its incorporation thereinto; and giving a flood irrigation, followed by proper tillage before sowing another crop in the herbicide-treated land.

If there has been a toxic build-up of residues because of the faulty application of soil-applied herbicides for 2 to 3 years, a wise step to enable such land to sustain all types of crops consists in growing certain crops which can remove large quantities of the herbicide residues from the soil. For example, maize can be grown both for fodder and grain, and another long-duration crop such as sugarcane, would also considerably reduce the amount of residues.

Place of Chemical Weed Control in Modern Farming

Multiple and relay-cropping patterns. The modern farming aims at growing a series of crops in multiple or relay-cropping patterns, the primary objective being to utilize the solar energy, which is a rich source of energy in this country, through the maximum production per unit of land, time and investment. The successful exploitation of this intensive and diversified agriculture depends upon growing improved high-yielding varieties of crop plants which respond to heavy doses of fertilizers and cupious watering. This type of farming is possible only where fertilizer is in plentiful supply. Water-supply is assured either through rains or irrigation, and pesticide use is a regular feature. Another facet of modern farming is related to those tracts which are subject to the vagaries of nature, such as drought and adverse soil conditions, and water-logging, salinity and alkalinity. In irrigated farming, unchecked weed growth will deprive the crops of the valuable inputs, such as nutrients and water, leading to the inefficient exploitation of their high-yielding potential. In dryland agriculture, which covers over two-thirds of the cultivable area of the country, the already low and erratic production due to adverse soil and weather conditions goes down still further in the presence of weeds which can grow even under adverse conditions of soil and climate. Chemical weed control in irrigation farming is the only way of making efficient and rational use of the costly inputs used in this type of farming. In dryland agriculture, chemical weed control ensures the stabilization of production in addition to a more efficient use of water by the crop plants.

Minimal cultivation. The concept of minimal cultivation, which is a valuable principle in both irrigated and dryland agriculture, ensures that weeds do not establish themselves securely in arableland. The raising of crops in multiple or relay=cropping sequence does not leave much time for carrying out cultivation. The control of weeds in this situation is possible through herbicide use. In the use of herbicides in such cropping patterns or sequences, it is very important to understand all the possible effects of these chemicals on the long-term effects of herbicides on soil, especially on the microbes which govern soil fertility to a great extent.

Biological methods. In biological weed control, a "natural enemy" of the weed plants is used. The requirement for the success of this method are: *(i)* the weed species must have been introduced and in the process of introduction, must have been freed from its natural parasites or predators, *(ii)* the natural predators or parasites must be introduced to prey upon or parasitize the weeds but they, in turn, must have been freed from parasitesin order to carry out their work of destruction unhampered, and *(iii)* the destructive agent must be highly specialized, so that it is unable to thrive even under starvation conditions on agricultural plants of the new habitat.

Root-borders, stem-borders and internal seed or fruit feeders are more highly specialized than the foliage-feeders.

An outstanding example of biological control of plants is that of prickly pear (*Opuntid* spp.) in Australia. Other noteworthy examples are the destruction of *Lantana* in Hawaiian islands, and St. Johnswort (*Hypericum perforatum*) in America. Attempts are under, way for the biological control of nutgrass in Hawaii, the common blackberry (*Rubus fruticosus*) in New Zealand. *Xanthium strumarium* (a weed poisonous to hogs) in Austalia, *Halogeton glomeratus* (a poisonous weed because of a high oxalate content) in south-west America. A lot of thinking is now being done for the control of water hyacinth (*Eichhornia crassipes*) through biological methods.

Somewhat different, but an interesting example of biological control is the destruction of aquatic vegetation in ditches by using herbivorous fish. The Hawaiian Sugar-Planters' Association reports that *Tilapia mosambica* of the sunfish family, by eating the vegetation and burrowing in the bottom of ditches to form spawing nests, destroys the roots of aquatic weeds and kills them. This method appears to have an additional merit, i.e. besides controlling submerged aquatic weeds, it can be used as human food. In Russia, this method is being exploited on a large scale. Biological control through "grass carp" which feed on aquatic weeds may be the safest, cheapest and the most effective of the methods of dealing with water vegetation.

The use of geese to control weeds in cotton is popular in some areas in southwest America. The young geese selectively feed on Johnson grass(*Sorghum halepenese*) without harming cotton.

The eradication of prickly pear by using the cochineal insect (*Dactylopius tomentosus* L.) in Maharashtra and Tamil Nadu is the best example of biological control in India. At the Indian Fisheries Laboratory, Bhopal, it was observed that grass carp (*Ctenopharyngodon idella*) controls effectively the submerged water vegetation comprising *Najasminor, Hydrilla, Verticillata, Ceratophylum demersum, Lemna minor* etc. Claims are also made in India of the control of *kans* grass (*Saccharum spontaneum*) by growing basket grass, the roots of which are supposed to excrete some substances inhibitory to *kans.*

Integrated Weed Control

In nature, a balance has been struck among all the components, both living(biotic) and non-living (abiotic) of an environment (habitat, system or region). This ecosystem concept has much relevance in that all the components coexist in perfect union, each component satisfying its own niche. The introduction of a foreign element,for e.g. a herbicide, may create some serious upheavals in the ecosystem. Further, the continued use of a herbicide may eliminate all the susceptible weed species and its place may be taken over by some resistant ones,or the existing ones may develop resistance. All these eventualities have to be borne in mind when recourse is taken to chemical weed control. One of the ways to avert such a situation consists in using low doses of herbicides, in rotating herbicides (using one particular herbicide for one year, missing it the next year or using some other herbicide) just like a crop-rotation and alternating intelligently chemical and non-chemical methods of weed control. An integrated weed-control approach is a sure way of minimizing the residue problems in plant, soil, air and water. An integrated weed-control involves the utilization of a combination of physical, chemical and cropping methods of weed control in a well-planned sequence so designed as not to affect the ecosystem. The nature and intensity of the species to be controlled, the sequence of crops that are raised in the rotation, the standard of crop husbandry, the ready and timely availability of any method and the economics of the different weed-control methods are some of thr potent considerations that determine the successful exploitation of the integrated weed-control approach.

Weed Control in Vegetable Cultivation

Weeds in vegetable cultivation fields are in different size, form and behaviour. They belong to many families varying in physiology, morphology and habits of growth. They compete with vegetable crops for water, nutrients, light and space and thus reduce yields. They also harbour insects and diseases. Weeds with larger and coarse leaves shaded vegetables and thereby restrict photosynthesis. Weeds are generally more vigorous in growth and absorb more mineral nutrients. Weed seeds germinate much earlier even under adverse conditions. They increase the expenditure on labours, equipments and chemicals for their control. Many weeds harbour fungus and bacterial diseases and insect pests. If the growth habits are similar the competition is very keen. The tillage operations are done constantly to remove the competition done by weeds. Weeds in vegetable fields may be annual, biennials or perennials.

The weeds in vegetable fields are generally removed by mechanical means such as cultivation with plough or tractor, mowing, burning etc. or by regular cultural methods. Each of these has advantages and disadvantages. With the discovery of herbicides the weeds are now controlled by chemical method. In home gardening weeds are usually removed with the help of spade, khurpi and hand pulling. In fact, hand weeding is the most efficient method. But due to high wages of labour all the manual methods are costly and time consuming. In large scale cultivation of weeds are controlled by tillage operations using various kinds of plough, harrow and cultivator.

The chemical method to control weeds is easier, less time consuming and less costly in comparison to mechanical method. The object of chemical weed control is to eliminate hand weeding and not to replace for more effective control of weeds.

The chemicals used to kill weeds and non-selective herbicides. The selective herbicides kill certain weeds but not the crops. Most of herbicides used in vegetables are selective. However, before use one must know the mode of action and nature of tolerance of a particular herbicide. The amount of herbicide a particular vegetable will tolerate with no damage varies with season climate and soil type. Herbicides in vegetables must be applied very carefully as it is applied as spray. The rate of chemical is in terms of active material; therefore water used has no importance.

The application of weedicides on vegetables in India has not expanded very much as because much research on this aspect has not been done in different soil types, climate and situations. The kind of weeds and population also varies in different states and areas. What kind and how much herbicide should be used for different kinds of vegetables in recommendations. Generally in vegetable cultivation herbicide is applied or immediately after sowing before it has emerged as pre-sowing treatments or as pre-emergence treatments.

The contact pre-emergence spraying is done to kill seedling weeds that have emerged before the crop. Pre-sowing treatments are done when a seed bed is prepared but for some reasons sowing is delayed.

In such case usually a non-persistent herbicide is used to remove weeds before sowing or transplanting. As pre-emergence spraying the chemicals like Diquat, Paraquat, Dimexan, Simazine and many other similar products are used. Dalapon is very effective to kill grass weeds. As post –emergence MCPB – salt, Dinosebamine, Barban, 2-4 –DB, MCPB are quite effective. The manufacturer's recommendations regarding quantity to be used, residual effect etc. should be followed.

There are other chemicals for use as pre-emergence spray like IPC, Dinitro amine, Chloro IPC, TCA, Karmax etc. In root crops, Aromatic oils are quite effective. There are many trade names with various formulations. Weed control methods are grouped into cultural, physical, chemical and biological. A combination of these methods gives effective and economic control than a single method.

Cultural Methods

Cultural methods, alone cannot control weeds, but help in reducing weed population. They should, therefore, be used in combination with other methods. Several cultural practices like tillage, planting, fertiliser application, irrigation etc., are employed for creating favourable condition for the crop.

If used properly, help in controlling weeds. In cultural methods, tillage, fertiliser application and irrigation, selection of variety, time of sowing, cropping system, cleanliness of the farm etc.

Field Preparation

Flowering of weeds should not be allowed. This helps in prevention of build up of weed seed population in the fields. Irrigation channels are the important sources of spreading weed seeds. It is essential, therefore, to keep irrigation channels clean.

Deep ploughing in summer exposes underground parts likes rhizomes and tubers of perennial and abnoxious weeds to scorching summer sun and kills them.

Planting Method

Sowing of clean crop seeds without weed seeds should be done. It is a preventive method against introduction of weeds. Sowing are taken up one to three days after rainfall or irrigation depending on soil type. Weeds already present in the soil start germinating within two or three days. Sowing operation with seed drill removes some of the germinating weeds. During this process, all the surface soil to a depth of 2 to 3 cm is disturbed uprooting germinating seeds.

Varieties

Weeds continue to germinate for long time in dwarf varieties resulting in high weed growth.

Planting Density

Plants of one type do not generally allow germination of other plants near their vicinity. It may be due to allelopathy and competition for growth factors. Closer planting of crops suppresses germination and growth of weeds.

Fertiliser Application

Crops like sorghum, maize, pearl millet and rice grow at a faster rate when nitrogenous fertilisers are applied and cover the soil earlier. Weeds like *Cynodon dactylon, Cyperus rotundus* do not respond to nitrogen application and they are suppressed by fast growing crops. If the weeds present are also responsive to nitrogen, then this method is not suitable.

Irrigation and Drainage

Frequent irrigation or rain during initial stage of crop growth induces several flushes of weeds. Under submerged conditions, tubers of *Cyperus* and rhizomes of *Cynodon* are killed due to lack of oxygen. Therefore, a good method of reducing *Cyperus rotundus* is by converting the field to rice cultivation.

Cropping Systems

The possibilities of a certain weed species or group of species occurring is greater if the same crop is grown year after year. Crop rotation can eliminate or atleast reduce difficult weed problems.

Weed management practices:

1. Follow recommended agronomic practices for land preparation, stubble management, seed bed preparation, timely sowing etc. so as to have a desirable crop stand to start with.
2. The crop should be maintained weed free atleast 8-9 weeks after sowing till canopy starts closing in by resorting timely interculture and hand weedings.
3. A hoeing in between crop rows is given 15-20 days after emergence of cotton seedlings to control primary perennials weed vegetation.
4. For control of weeds, herbicides such as a pendemethalin, alachlor (1.5 kg/ha) and diuron (0.5 kg/ha) may be applied as pre-emergence and fluchloralin (0.75 kg/ha) as pre-plant incorporation.

Weed Control Challenges

Organic vegetable production is a method that does not allow synthetic fertilizers, crop protectants, or genetically modified seed. Growers cannot just substitute "organic" products, but must create a long-term farming system with healthy soil, crop rotations, cover crops, and preventative pest control via cultural, physical and mechanical means.

Weed control in organic vegetable production can be challenging. Weeds compete for light, water, and nutrients. Sometimes, they also encourage insect and disease problems. Short-term control is the seasonal control of weed pests, but long-term control is also possible through depletion of the weed seed bank in the soil.

First things first, weeds must be correctly identified in order for control measures to be appropriate. They can be classified based on whether they are grasses or broad leaves. Weeds also can be classified based on their lifecycle. Winter annual weeds germinate in the fall, overwinter, then re-grow in the early spring and generally die back by summer.

Summer annuals are one of vegetable growers' biggest headaches, germinating in the spring and early summer and growing along with the crop. These are weeds such as lambsquarters, pigweeds, nightshades, and velvetleaf. Biennials, such as Queen Anne's Lace, germinate and grow vegetatively the first year, overwinter, then flower and produce seed their second year.

Quackgrass, nutsedge, Canada thistle, and milkweed are perennial weeds that grow indefinitely and propagate not only by seed but by underground structures such as nutlets, rhizomes, etc.

The Practice of Rotation

Crop rotation is one of the most essential practices of organic vegetable growers. Because crops compete differently with weeds, rotation is an important aspect of weed, insect, and disease control. It is often difficult to plan for with vegetable crops since size of plantings may vary. Growers need to know crop families (brassicas, solanaceous, umbelliferaceae, cucurbits, legumes, sweet corn, alliums, lettuces), and should rotate by crop family. Try to design a plan for your fields for the next three to five years.

Some recommendations to follow include:

- Rotate back to a family of crops only once every three to five years.
- Have a portion of the farm fallow at any given time.
- Plan for cover crop rotations.
- Plan for double cropping, if possible.
- Take into account weed competitiveness.

For example, carrots, parsley, and lettuces do not compete well with weeds and are delicate. Put in these types of crops after a number of years of crops that can be cultivated and compete well.

Cover crops can be an effective method of weed control, since fast-growing, thickly sown crops will out-compete weeds. Choose a weed-controlling cover crop based on your soil needs and the type of weeds prevalent in the field.

Legume cover crops such as clover or vetches add nitrogen; non-legumes (rye, wheat) add organic matter to the soil. Some cover crops, like rye, are allelopathic, meaning they give off substances that inhibit weeds. Cover crops should be planned for in a rotation plan.

The Mulch Advantage

Both plastic and organic mulches, such as straw, can have many advantages in addition to acting as a physical barrier to weeds in the crop rows. When using black plastic, be sure to prepare your field appropriately and have enough soil moisture so that the mulch is applied smoothly and has good contact with the soil. Disadvantages to plastic mulch include weeds in between rows and in the planting holes. Disposal is also a problem, and if possible, used mulch should be taken to a recycling facility.

Mowing weeds is a temporary solution, but can prevent them from actively competing with the crop and from going to seed. Propane flamers can be very effective in killing small broadleaf weeds. Growers can use handheld flame weeders as well as larger, tractor pulled units for use on larger areas.

And finally, one of the most important techniques in organic weed control is cultivation. Effective cultivation includes knowledge of the crop, the weed, and the specific implement used. There are many different tools and timings of cultivation to bury or uproot the weed. Initial tillage prepares the field for planting and removes any growing weeds.

The stale seedbed technique requires that after initial preparation, the field is watered and weeds are allowed to germinate. The field is then tilled shallowly (or flamed) to remove weeds.

This is repeated two or three times to deplete the weed seed bank in the germination zone. Tools, such as tine weeders and various harrows, are useful for the stale seedbed technique.

Edge of the plastic, between row and in-row cultivation for organic weed control, is both an art and a science. Knowing the right timing for the weed and the soil conditions is very helpful but not always possible given the weather. But choosing the proper equipment, setting it correctly, and driving skillfully are all up to the operator. Traditional tools such as field cultivators are useful, but there are many other choices available including basket weeders, spyders, and finger weeders.

In conclusion, organic weed control can be achieved by an integrated program of cover crops, crop rotations, mulches, and mechanical controls such as flaming, mowing, and most importantly, cultivation. Preventing weeds from going to seed will deplete the weed seed bank and eventually lead to lower weed pressure and cleaner fields.

Biological Control of Weeds

Plants that have become weeds in Australia are rarely invasive and troublesome in their home country. This is often because populations in the home country are regulated by a variety of natural enemies such as insects and pathogens (disease-causing organisms like fungi and bacteria) that attack the seeds, leaves, stems and roots of the plant.

If plants are introduced to a new country without these natural enemies, their populations may grow unchecked to the point where they become so prevalent that they are regarded as weeds.

What is Biological Control?

The biological control approach makes use of the invasive plant's naturally occurring enemies, to help reduce the invasive plant's impact on agriculture and the environment. It simply aims to reunite weeds with their natural enemies and achieve sustainable weed control.

Biological control is rarely a 'silver bullet', as not all weeds have biological control agents that would be considered safe for introduction in Australia.

These natural enemies of weeds are often referred to as biological control agents. It is critical that the biological control agents do not become pests themselves. Considerable host-specificity testing is done prior to the release of biological control agents to ensure they will not pose a threat to non-target species such as native and agricultural plants.

Not all weeds are suitable for biological control. Developing a biological control project requires a substantial investment, sometimes costing millions of dollars, from stakeholders.

A biological control agent is generally only used when the cost of conventional control methods such as herbicides, mechanical control or fire is so great, both in dollar terms and impact on the environment, that there is little option than to pursue the biological control avenue.

CSIRO has been working on the biological control of weeds since the 1920s, starting with the biological control of prickly pear.

CSIRO now has many active biological control projects underway for both temperate and tropical Australian weeds which cause problems in natural, pastoral and agricultural ecosystems.

The Process

A weed becomes a problem in the introduced range because its population density fluctuates around an equilibrium that is above a threshold at which the weed begins to affect the economic or ecological sustainability of the ecosystem.

Best case scenario of the course of events in classical biological control programs targeting weeds invading habitats such as rangeland, pasture, or natural ecosystem. Following their introduction and establishment, populations of biological control agents build up to very high levels due to the abundance of their host plant. Eventually their attack on the plant causes a decline in the weed population.

This, in turn, leads to a decline in the numbers of biological control agents until an equilibrium is reached between the amount of damage caused by an agent and regeneration by the weed. In a successful biological control program this new equilibrium is below the damage threshold that the ecosystem can tolerate.

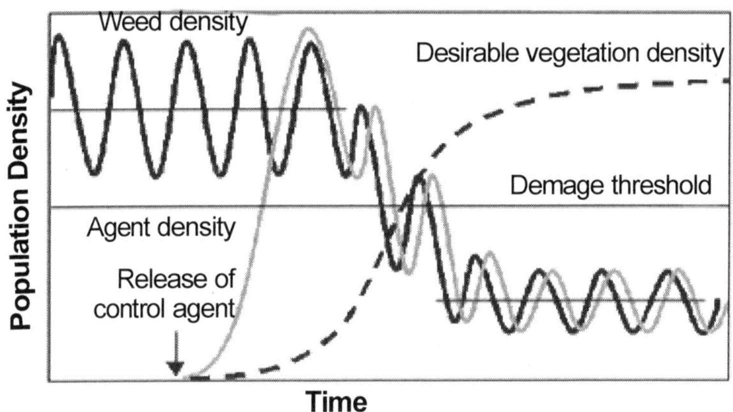

Biological control of weeds is the deliberate use of natural enemies to reduce the density of a purticular weed to a tolerable level The objective of biological weed control is not eradication but simply the reduction of the weed population to an economically low level In fact for biological control to be continuously successful, small numbers of the weed host must always be present to assured the survival of the natural enemy.

The two most frequently cited examples of successful biological weed control are the destruction of the prickly pear cacti (Opuntia; spp.) in Australia by an imported moth [Cactoblastis cactorum) and the control of St. Johnswort Hypericum perforatum) millepertius perforé) on rangeland in California, Oregon, Washington, and British

Columbia. These examples demonstrate that biological control can provide a permanent solution to serious weed problems. The first insects were introduced 50 years ago and the Australian prickly pear is still under control. St. Johnswort is presently controlled at two per cent of its former density in British Columbia.

Weeds are plants growing where they are not wanted" and most of our serious weeds have been introduced from other parts of the world. Of the 107 noxious Canadian weeds, 78 have been introduced from Europe or Asia. One of the reasons why these plants are so noxious in their new habitat is that their natural enemies are often absent. Biological control has most frequently been applied against these alien weeds and attempts are made to restore the natural control of these weed pests by introducing one or more host-specific, damaging natural enemies from the native region of the weed.

Methods Used in Biocontrol of Weeds

The classical approach to biological control of weeds involves the introduction of host-specific natural enemies of allien weeds. Recently the approaches utilized in biocontrol programs have been expanded to include two other methods: 1. augmentation of natural enemy populations; 2. application of "biological herbicides".

Augmentation includes the periodic release and/or distribution of natural enemies. Work is presently being conducted in the United States to determine the effect of mass rearing of a native moth Bactra verutana) in the laboratory and releasing it in fields of yellow nutsedge Cyperus esculentus) (souchet comestible).

Biological Control Agents

Insects have been most frequently used as biological control agents of weeds and this will likely continue. The reasons are that there have been major successes using phytophagous insects and almost all of the scientists working in biocontrol of weeds are entomologists. However, recent research has demonstrated the potential of other organisms, including plant pathogens, nematodes, and fish.

Procedures in Classical Biological Control

The first step in a biological weed control program is to determine the suitability of the weed for this approach. Not all weeds are suitable and those with the following characteristics are generally least suited for biological control:

1. Weed species which are valued in other situations are not good candidates for this approach. for example, blue weed Echlum Vulgate) (vipérine) is a serious pasture weed, but it is also a desirable honey plant. As the biocontrol agent cannot be limited in area, like chemical or mechanical treatments, 'weeds' that are of values in other situations are not suitable targets for biocontrol.
2. Weeds that are closely related to economic crops are not good candidates for this method. The closer the relationship the less possibility there is that a biotic agent could distinguish between the weed and the crop. For example, no insects have been found that attack wild oats Avena fatua) (folle avoine) and do not attack cultivated oats and our other cereals.
3. Native weed species are not generally amenable to this approach. However, the native range of an introduced allien weed provides a source from which a parasite can be introduced.
4. Weeds of cropland under intensive cultivation are generally not suited to this approach. Since the biological control agent is specific to only one weed species, little would be gained if one weed, such as lambsquarters, was controlled in the corn crop as numerous other weed species would soon occupy the available space. However, biocontrol is particularity suited to rangeland situations where a single, dominate weed species is troublesome.
5. Minor weed problems are not generally suited to this approach. The target weed should infest large areas. Since there are only four scientists in Canada working more or less full time in biological weed control, minor weeds usually cannot be considered.
6. If eradication of the weed is desired (e.g., poisonous weeds), the method is generally not applicable.

Therefore, in classical biocontrol of weeds, the ideal target weed is an aggressive, introduced weed which infests large areas of marginal land such as rangeland, pastures, and waste areas.

The second step in the program is to conduct a survey overseas to determine if there are any parasites available for introduction against the weed. A survey is also conducted in Canada to ensure that a prospective biotic agent is not already present in Canada and to determine what parts of the target weed are not attacked in Canadian populations.

For example Canada thistle Cirsium arvense; (chardon des champs) is attacked by a number of seed feeders in Canada and, therefore, it would be of little value to introduce other organisms that attack the seed heads. Surveys are being conducted in Europe on the following weeds:

English name	***French name***	***Scientific name***
Bindweed	Liseron	Convolvvulus spp.
Absinthe	Armoire	Artemisia absinthium
Narrow leaved hawk's beard	Crépis des toits	Crepis tectorum
White cockle	Lychnide blanche	Lynchis alba
Bladder campion	Silene enflée	Silene cucubalus
European buckthorn	Nerpun commun	Rhamnus cathartica

The third step is to determine the potential effectiveness of the parasite in Canada In an attempt to eliminate ineffective agents before importation and screening tests in Canada. Insects that have been collected in certain parts of Europe may not survive under Canadian conditions.

The next and perhaps most important phase of the program is to determine the safety of the selected parasite for release in Canada. The introduced agent must not attack or damage any desirable plants in Canada. Very extensive tests are conducted in the quarantine facilities at the Agriculture

Canada Research Station at Regina, Saskatchewan. These studies involve thorough investigations of the agent's biology and its host range to demonstrate that the introduced agent is host-specific and will not become a pest of an economic crop. Some insects that are presently being screened at the Regina laboratory include:

Liothrips spp.—a sucking insect which attacks ragweed (Ambrosia artemisiifolia) (petite herbe à poux). Tephritis dilacerata—a seed-head fly attacking perennial sowthistle (Sonchus arvensis) (laiteron des champs). Argyroploce striana—a root moth attacking dandelion (Taraxacum officinale) (pissenlit).

If it is found during these tests that the candidate biological control agent is not host-specific, the organism is destroyed and the project is terminated.

After the screening tests are complete, a report is prepared for joint approval by the Canadian and U.S. governments. A report on the nematode, Paranguina picridis, which forms galls on Russian knapweed [Acroptilon (Centaurea) repens] (centaurée de Russie), was recently submitted and permission for limited field trials has been granted.

Once the agent has been determined safe and permission to release it has been given, the fifth step in the program is to establish the biocontrol agent in Canada on infestations of the target weed.

This part of the program requires the co-operation of farmers. Infestations of the target weed need to be located and the release site must be maintained for some time with minimal disturbance. This stage is the most critical in the program as many agents have failed to become established because they were not adequately cared for during the initial phases of establishment in their new environment. Attempts are being made to establish insect agents on the following weeds in Canada:

Nodding thistle (chardon penché) Carduus nutans	Rhinocyllus conicus (seed-head beetle)
	Ceutorhynchidis horridis (root weevil)
Canada thistle (chardon des champs)(Cirsium arvense)	Urophora cardui (stem gall fly)
Bull thistle (chardon vulgaire)	Urophora stylata (seed-head fly)
Russian thistle (soude roulante) (Salsola pestifer)	Coleophora parthenica (stem mining moth)
Diffuse and Spotted knapweed (centaurée diffuse et centaurée maculée) (Centaurea diffusa, C.Maculosa)	Urophora affinis and U.quadrifasciata (seed-head flies)

Once the insect agent is established, the step is to determine if the agent is increasing and its effect on the weed. The beetle on nodding thistle and the seed-head flies on the knapweeds are well established and are rapidly increasing.

Two major projects can be considered complete in Canada. As mentioned earlier the population of St. Johnswort in British Columbia has been reduced to two per cent of its former density by two leaf-feeding Chrysolina beetles. Tansy ragwort Senecio jacobaea) (séneçon jacobée) has been successfully controlled in Nova Scotia by the defoliating cinnabar moth (Tyria jacobaeae).

From the above examples it can be seen that biological weed control programs have been primarily concerned with introduced insects as biotic agents and the approach has been directed against aggressive, alien weeds in rangeland situations. However, recent research results with other types of biocontrol agents, particularly plant pathogens, has extended the application of biological control to aquatic weeds and those of cultivated land.

Plant Pathogens As Biocontrol Agents of Weeds

Plant pathogens offer two advantages over insects as biocontrol agents of weeds: 1) they are often more host specific, and 2) they can be applied with conventional spray equipment at a time when the weed is at its most susceptible stage.

Conventional Approach

As discussed above, biological control of weeds has conventionally been applied against alien weeds by introducing one or more of their natural enemies. Control of a weed with a plant pathogen used in this manner was achieved against skeleton weed (Chondrilla juncea), the major weed of wheat in Australia. A host-specific rust, Puccinia chondrillina, was collected in Italy and introduced into Australia in 1971. The rust spread very rapidly and at present controls the weed over most of its range. Control has been so spectacular that herbicides are no longer being used to control this weed in Australia.

Numerous introduced weeds in Canada are known to be attacked by plant pathogens in their native ranges. However, research in this area has not yet been established in Canada.

Biological Herbicides

A new approach to biocontrol of weeds has recently been pioneered in Arkansas with the application of a spore suspension of an endemic fungus. The fungus controlled 99 per cent of a serious weed (Northern jointvetch) in rice, without damage to the rice crop. The disease was already present in the fields but did not damage its host because it normally attacked too late in the season. However, when the fungus was applied early (in the seedling stage), it destroyed the weed. The fungus is now being tested on a field scale and there appears to be no biological or technological limitations that would prevent commercialization of the fungus as a' mycoherbicide'.

Possible targets for this approach in Canada are numerous as practically all Canadian weeds have plant pathogens recorded on them. for example, wild oats (Avena fatua) (folle avoine) and other weeds of our cultivated land can be controlled to some extent by cultural and mechanical means. However, they still remain as serious contaminants of our crops. Wild oats is attacked by numerous plant pathogens which normally attack late in the season and do not appreciably damage the weed. If host-specific pathogens of wild oats were applied early in the season, some degree of control could possibly be achieved.

An attractive feature of this method is that it does not involve the introduction of any new organism into our environment but merely the use of host-specific pathogens at a time when they are most effective. This method may also prove to be valuable for integrated weed control systems. For example, most weeds are controlled in corn with standard herbicide application, but a few weeds such as quackgrass (Agropyron repens) (chiendent) and yellow nutsedge (Cyperus esculentus (souchet comestible) are more difficult to control. The application of host-specific pathogens could be integrated into the present weed control system to control these types of weeds.

Aquatic Weed Control

Biocontrol of water weeds is being investigated in Florida and other parts of the U.S. The possibilities of using insects, plant pathogens, and a fish, the white amur (Ctenopharynogondon idella) are being studied. Water weeds are also a serious problem in many areas of Canada and the developments in biocontrol are being carefully observed. The disadvantages of using herbicides in water systems are numerous and alternative methods should receive high priority. There is considerable controversy concerning the use of the fish for biological control. Research is continuing and further studies should demonstrate that the advantages will by far outweigh any possible disadvantages.

Biological Weed Control Program at Macdonald College

The research in biological weed control at Macdonald College is composed of two segments: 1. The release of screened biotic agents on infestations of weeds in Quebec; 2. Investigation of the potential of native natural enemies of certain weeds in Quebec.

Release of Biotic Agents

Insects that have been screened and are available for release are shipped to Macdonald College from the Regina laboratory. These insects are then taken to desirable field release sites and liberated on the target weed. This program has involved the co-operation of numerous Quebec farmers and its continued success depends on their support. Last year, insects were released on the following weeds in Quebec :

Canada thistle (chardon des champs) (Cirsium arvense)	Urophora cardui (stem gall fly)
Bull thistle (chardon vulgaire) (Cirsium vulgare)	Urophora stylata (seed-head fly)
Leafy spurge (euphorbe ésule) (Euphorbia esula)	Hyles euphorbiae (defoliating moth)

Additional releases of these insects will be made this year and releases of Chrysolina beetles on St. Johnswort, Tyria.Jacobeae (cinnabar moth] on tansy ragwort and Uronhora spp on spotted knapweed infestations in Quebec are planned for this year.

Urophora cardui has previously been released at St. Hyacinthe and Rhinocyllus conicus has previously been established on nodding thistle in the Lac St. Jean region.

Native Natural Enemies

This program involves surveying of weed populations, collection of natural enemies (plant pathogens and insects) and determination of the potential of selected organisms as biocontrol agents. At present, two postgraduate students, Jean-Guy Champagne and Harry Hartmann, are investigating the host-parasite relationships of Canada thistle and its rust pathogen and ragweed and its natural enemies. In addition endemic plant pathogens of other weeds, such as quackgrass and yellow nutsedge, are also being collected and studied.

Summary

Biological weed control has recently received renewed interest because it is an environmentally compatible method of weed control without residue and pollution problems. However, biological weed control has its weaknesses and is not suitable for all weed problems. Indeed, only a few of our many noxious weeds have been investigated for biocontrol and so far only two are sufficiently advanced to be considered complete successes. Undoubtedly, with continued interest and research support, the general assumption that biological weed control is used only as a last resort where chemical and other methods have failed will be proven incorrect.

Classical Biological Control of Weeds

Around the turn of the century, a European plant that is somewhat toxic to cattle and sheep was first reported in northern California near the Klamath River. It increased and spread rapidly, so that by 1944 "Klamath weed" was found on over two million acres of range land in thirty counties. Land values decreased so much because of the presence of this weed that ranchers couldn't get bank loans for improvements. Herbicides to control the weed were costly and much of the infested land was inaccessible for spraying anyway. The idea of controlling Klamath weed by importing insects that fed on it was proposed as early as 1922, but authorization was not obtained until 1945. By then it was impossible to consider importations from Europe because of World War II. Fortunately, Australian scientists had been working on biological control of the same weed problem in Australia since 1929, so it was easy to import three species of beetles (originally from England and Europe) that had shown promise for controlling Klamath weed in Australia. One of the three beetle species did not survive, but the other two made it through all the tests in quarantine. One was released in the spring of 1945 and the other in February 1946. Both quickly became established; their offspring were then distributed throughout infested areas. More than 3 million beetles were collected for redistribution in California in 1950 from one colony where 5000 beetles were released in 1945-46! Beetles were also successfully established in Washington, Oregon, Idaho and Montana. Within ten years after the first release, Klamath weed was just an uncommon roadside weed in California. Land values increased, and cattle and sheep were no longer in danger. It was estimated that the beetles saved the California agriculture industry about $3.5 million dollars per year in 1953-59, and that these saving continue to accrue each year.

The story of Klamath weed is just one example of successful classical biological weed control. Many of the other success stories of biological weed control until about 1960 were also on alien perennial species infesting grazing lands, but more recently annual and biennial plants, aquatic and semi-aquatic weeds, and native weeds have been the target of biological control research, too. Biological control of weeds historically has usually been considered only after all other management options have been unsuccessful, often decades after the weed is discovered. By that time, the weed can cover immense areas with dense, intractable populations, and there can be thousands of seeds per square foot. Globally thousands of plant species have been accidentally or deliberately introduced into areas outside their natural distribution without their co-evolved natural enemies. Despite sophisticated attempts to exclude potentially weedy species, new species are introduced each year. Some of these introduced plant species will establish and become weeds of agriculture, forestry, pasture, urban situations, or natural areas. With increasing restriction on use of herbicides, classical biological control is often the only choice for managing these weeds in a sustainable, environmentally-compatible manner. Biological control is not a panacea, but is one of the options that should be considered when faced with a new weed problem. When exclusion, eradication, or a quick kill of a weed population is the goal, classical biological control is generally not applicable because the host must be present, classical biological control is relatively slow-acting, and results cannot be guaranteed.

The terminology of biological control of weeds is similar to that used in biological control of other pests, but the terms are sometimes used in slightly different ways. Classical, or inoculative, biological weed control was recognized and practiced first, with the first modern use occurring in the late 1800's. It remains the most-often practiced type of biological weed control, and is generally conducted by government agencies over large areas. Classical biological weed control relies on establishment by "inoculating" relatively small numbers of individual agents at release sites, from which agents spread on their own. Classical biological weed control is thus essentially an ecological approach to pest management.

Arthropods, nematodes, vertebrates, and increasingly, microbial species, are used as weed biological control agents. It is important to use the most appropriate type of agent to match the needs of the program. Classical biological control is limited to importation and release of host-specific natural enemies, and often works best in relatively stable systems, such as pastures. The major risks are known and evaluated before agents are released in the environment. If the target weed is closely related to a crop species, the number of usable natural enemies will be greatly reduced. Successful programs result in cost-efficient, sustainable management of the target species with no or minimal non-target effects.

Augmentative, or inundative, biological weed control was recognized as a possibility about the same time as classical biological control, but was not used widely until much later. Technology is developed for producing relatively large numbers of agents and "inundating" populations of the target weed in the field at a time when it is most susceptible to attack. Commercial formulations of microbial agents used in this way are known as "mycoherbicides", which is an extremely promising method.

Most augmentative biological control of weed agents do not become established after release, so must be reapplied each season. Thus, augmentative release is essentially technological in nature. Conservation and broad spectrum biological weed control are more recent developments. They are essentially ecological and technological approaches, respectively.

Although all types of biological weed control should be considered when developing an integrated management program for a particular weed, only classical biological weed control.

Once a particular weed is targeted, there are several steps that are generally followed in a classical biological weed control program. During the "pre-release" or "pre-introduction" (of the biological control agents) stage, extensive information is gathered on the weed and all its known natural enemies. This information is submitted to appropriate State and Federal agencies to obtain approval for the project, and then foreign exploration for good natural enemies can begin. Candidates for introduction as control agents are studied extensively, along with the ecology of the target weed. Promising agents are tested for host-specificity, generally under artificial conditions in laboratories, greenhouses, cages, or field plots in the weed's native area. Once these exhaustive studies have been completed, the data is submitted to the USDA's Animal and Plant Health Inspection Service (APHIS) and the Environmental Protection Agency, who share responsibility for approving weed biological control agents. The last step in the pre-release stage is to import the approved natural enemies to a quarantine facility where they are reared through one or more generations to be sure they are not diseased or have any of their own parasites. In the "post-release" or "post-introduction" stages, the natural enemies are massed-reared and released in the field. Developing the mass-rearing techniques is often one of the most difficult parts of the whole operation, and sometimes the agents can only be distributed in very small numbers at first. (This is also why classical biological control researchers often refer to the method as "inoculative" because even with the best rearing methods, often there are only a few hundred individuals available for release.) Once the natural enemies are in the field, they should be monitored to determine if they become established and if they are having any impact on the target weed population. Even though this is an important step in the program, it is often neglected because funds are frequently available only for discovery and release of agents, but not for evaluation and monitoring. Finally, each natural enemy species is redistributed to other areas in the target weed's distribution, if spontaneous self-dissemination is not adequate.

There have been several recent policy developments that should result in increased support for biological weed control programs in the coming years. In 1990 the National Biological Control Institute was established as part of USDA's APHIS to provided leadership in the area of biological control. In 1992 USDA's APHIS announced a "Biological Control Philosophy" that stated support for biological control as the base strategy of integrated pest management and indicated the agency would "develop regulations that facilitate the release of safe biological control agents." This policy requires fundamental changes that are now being implemented. The USDA Forest Service (FS) and Department of Interior announced major policy changes to "ecosystem management", and in 1993 USDA FS established the National Centre for Forest Health Management in Morgantown, WV. The Centre has established biological control as one of the key strategies in their program development. In June 1993 the Clinton Administration announced a major pest management policy change: the US will reduce pesticide use by increasing biological and cultural controls. In addition, several major reports have been released recently highlighting the importance of a shift to biological controls from pesticide use, and a "biological control philosophy" has been espoused by many organizations worldwide.

Biological Control 'Silver Bullet'

Biological control is more often associated with pest insects but there have been a number of examples where beneficial insects and pathogens have been deployed against weed species. In India, a rust pathogen, *Puccinia spegazzinii*, is being tested in a classical biological control (CBC) programme against the invasive weed *Mikania micrantha*, which originates from tropical America.

Mikania is an invasive alien weed that can smother both agroforestry and natural forest ecosystems, leading to invasions of home-gardens and plantations in many regions of tropical Asia. It was introduced to India as a cover crop and as camouflage for airfields in the 1940s, but its impact has escalated in recent decades due to widespread cutting and degradation of natural forests; Mikania, having colonised the degraded forestland, then invades adjoining tea gardens and village vegetable plots. The weed has also become a problem throughout the moist forest zones of Southeast Asia.

Control of Mikania has relied on slashing and herbicides but these methods are expensive yet not effective, as well as being environmentally damaging. Indeed, herbicides have been used at such high levels that residues have prejudiced sales of the tea crop. Therefore, a project to investigate an Integrated Pest Management (IPM) approach to Mikania control has been funded by DFID. This involves CABI Bioscience, UK and three Indian organisations: Kerala Forest Research Institute, Directorate of Biological Control, and Assam Agricultural University. The project has been to assess the socioeconomic impact of Mikania on the basis of mapping its distribution and monitoring its spread, and to evaluate potential biological control agents. Since no effective natural enemies of Mikania have been found in India, attention was turned to potential pathogens in the weed's native range in the Americas.

5

Weed Management of Horticulture Crops

Environment and sustainable development issues cut across the whole spectrum of economic activities, and particularly concern agriculture, forestry and fisheries. In spite of their diversity, Small Island developing States share common constraints that impede their efforts to develop the agriculture, forestry and fisheries sectors, while preserving their environment. Among the constraints are fragile ecosystems, vulnerability to natural hazards, population problems, poor communication facilities, and dependence on external economic factors. They run the added potential risk of rising sea level. On the other hand, most islands have the potential to diversify their economies for export and import substitution, for tourism and for drawing greater benefits from the fisheries resources of their exclusive economic zones. In determining their individual sustainable development objectives, Small Island developing States can draw great benefit from improved exchange of information with other islands sharing similar problems.

The agriculture, forestry and fisheries sectors have three vital functions:

- supplying food and raw materials;
- providing employment and income in rural areas; and
- rendering essential environmental services for present and future generations in the maintenance of a large part of the earth's renewable natural resources.

In most countries, these sectors no longer have the capacity to perform all three functions efficiently and simultaneously.

The issues presented in this paper include approaches to sustainable management of the resource base by the agriculture, forestry and fisheries sectors, enhancement of self-reliance through a combination of diversified and intensified primary production from renewable resources, generation of income from the tourist industry, and capacity-building to meet the challenges posed by sustainable agriculture and rural development.

Macro-economic Conditions

More than other countries, small island developing States are dependent on the international trading system for their livelihood, which makes them vulnerable to the vagaries of the world market. This characteristic of small island developing States means that the achievement of sustainable agriculture is more complex than for countries which have to rely less on the world market. The relationship of trade policies to sustainable development and the environment is currently the subject of intense debate.

Fitting into the global economy is especially hard for small island developing States, which face a number of drawbacks:

- limited population size making economies of scale and labour specialization impossible;
- out-migration of skilled personnel, resulting in a high degree of dependence on foreign capital, external assistance and expatriate technical assistance which may not always be sensitive to local conditions;
- a small monetized sector with local private investment limited to low-risk activities such as trade and services; a heavy burden on the public sector faced with limited domestic income-generating capacity reliant on sources such as import duties, fishing rights and remittances from nationals working abroad;
- a narrow resource base creating a dependence on imports of consumer and capital goods;
- and exports limited to primary products from agriculture, forestry, fisheries and mining prone to considerable fluctuations in output and price.

Sustainable development, including food security, nutritional well-being, poverty reduction, and rural development require inputs from many technical disciplines and should be addressed holistically in the planning process. This has been stressed in the structural adjustment programmes.

In the provision of policy advice and programming assistance, the main challenge is to assist small island developing States to incorporate environmental and sustainable development considerations into the mainstream of agricultural planning and policy, in line with Agenda 21 recommendations. Sustainable agriculture and rural development (SD) has three essential goals:

- food security through an appropriate and sustainable balance between self-sufficiency and self-reliance;
- employment and income generation in rural areas, particularly to eradicate poverty;
- natural resource conservation and, environmental protection.

"The International Cooperative Programme Framework for Sustainable Agriculture and Rural Development (ICPF/SARD)"

Since 1991, FAO has been streamlining its Special Action Programmes (SAPs) related to sustainable development. SAPs are clusters of projects which correspond to the priority areas indicated in the den Bosch Agenda for Action and subsequently incorporated into the programme areas of UNCED's Agenda 21, Chapter 14 on SARD, with the addition of two specific SAPs on forestry and fisheries as well as a third SAP directly related to the follow-up to the International Conference on Nutrition.

The component SAPs of the ICPF/SARD are as follows: SAPs focusing on policy advice and planning assistance: (i) Country Policy and Programming Assistance for Food Security and Sustainable Agriculture and Rural Development; (ii) Tropical Forests Action Programme; and (iii) Responsible and Sustainable-Fisheries; SAPs focusing on people's welfare: (i) Nutrition and Food Quality; (ii) People's Participation in Rural, Development; and, (iii) Sustainable Development of Rural Households; SAPs promoting the sustainable management of the natural resource base: (i) Land Conservation and Rehabilitation; (ii) Management of Rural Water Resources; (iii) Conservation, Development and Use of Plant Genetic Resources; and, (iv) Global Programme for Animal Genetic Resources; SAPs dealing with the sound use of agricultural inputs: (i) Integrated Plant Nutrition Systems; and (ii) Integrated Pest Management. Regional and national projects to implement the ICPF/SARD may be formulated for the small island developing States expressing interest.

It will take several years before the SARD process can demonstrate its environmental, social and economic benefits. During the transition period, heavy investments of labour, capital, technology and research will be required to accomplish SARD's goals.

To be successful, SARD must be perceived as a pivot around which the environmental, social and economic fabric of a country evolves. SARD requires a comprehensive approach where all essential elements are brought together within a single framework. Technical assistance and capacity-building should be pursued with all actors involved in agriculture, who have different but complementary roles.

At the national level, sustainable agricultural and rural development would integrate sectoral and cross-sectoral programming assistance in agriculture, forestry, fisheries, poverty alleviation and nutrition; at the community level, people's participation would be enhanced through human resources development and strengthening rural people's organizations; at the household level, all aspects of sustainable management of the resource base, appropriate technology, sound use of inputs and diversification of sources of income would be integrated to increase levels of production. Economic agents at all levels and in all sectors, including national governments, develop and carry out strategies for sustainable development and management of land, water and biotic resources. This is why all must have access to reliable information and be involved in the formulation and application of national strategies reflecting the recognized congruence of individual and society interest, both short and long-term. The values of conservation and rehabilitation need to be widely publicized and demonstrated, and technical advice and assistance provided to those needing help. Governments can create the conditions and institutions to encourage sustainable production practices by farmers, fisherfolk and other economic agents.

"Agricultural Sector Review in Malta"

In view of assisting the Government in adapting its agricultural policy to the Customs Union and eventual European Union membership, this project (1991-93) reviewed the challenges and opportunities facing the sector. Over the years, Malta has come to depend increasingly on imports of agricultural commodities, mainly from EU countries. This is due to decreasing land resources available for agriculture and rising food demand from the, tourist industry.

The, number of factors contributing to trading with EU countries include their proximity but also reduced import tariffs as compared to other countries, the EU policy regarding export subsidies to non-EU countries, the size of the market and facilities for handling and storage of imports. Besides focusing attention on functional improvements of the agricultural support services (notably experimental work, demonstration and extension), particular emphasis was placed on strengthening the capacity of the sector to retain the land base for productive and ecological purposes.

The measures recommended include: improvement of public sector management in agriculture and fisheries; improvement in the management of land, fresh water and marine resources; introduction of high-yielding, cost-effective and environmental friendly production techniques; improvement of marketing systems for crop, livestock and fisheries products; a review of the land tenure system; rationalization of trade measures to control imports of agricultural commodities; and a review of price control measures. Financial incentives could be required during the transition period to boost productivity and improve the incomes of full-time farmers.

Implications of the Uruguay Round Agreement

The Agreement on Agriculture of the recently concluded Uruguay Round will have wide-ranging implications for the developing countries, including the small island developing States. Its anticipated impact on small island developing States arises both from expected changes in agricultural commodities markets and from limitations on their freedom to implement policies adapted to local conditions. The former are likely to be contingent and temporary, whereas the latter are structural and long-lasting. Nearly all small island developing States are net food importers. Because of the reduction in agricultural support in the developed countries, the structural surpluses they generated in the past are likely to disappear. As a result, world market prices will likely rise, especially as developed country exporters are committed to reducing export subsidies. Several small island developing States already faced with serious balance-of-payments problems will find it difficult to maintain their food import volumes without sacrificing other essential imports.

The Agreement contains some safeguard provisions concerning the possible negative effects on least-developed and net food-importing countries. These cover areas such as food aid, technical assistance and short-term assistance in financing normal commercial imports.

Whether these provisions would constitute adequate coverage against higher import bills is not clear in view of certain conditionalities for eligibility. On the other hand, the prices of commodities exported by small island developing States are likely at best to rise only marginally as a result of the Agreement. Indeed, there may be a reduction in the export earnings of countries which have benefited in the past from preferential access to developed country markets. These countries may, however, gain from concessions in non-agricultural sectors. Overall, there is no certainty that they will be net gainers from the Uruguay Round.

The most significant implications of the Agreement relate to the policy environment. Policies allowed under the Agreement are those which have at best minimal production or trade distortion effects. Policies that do not meet this basic criterion may be kept but expenditures on them must be reduced. The least developed countries are not expected to cut support.

In small island developing States, maintaining or increasing food production in order to reduce dependence on imports is contingent upon substantial price support. The Agreement implies that this type of support will be more limited in the future. This has implications for minimum safeguards that they may want to establish (in the form of food security stocks or funds) to protect themselves against world market volatility. It is difficult to make even a broad evaluation of the Agreement on SARD. Policies that can help the environment are allowed under the Agreement, but these can only be implemented in a healthy overall economic climate and may be too costly for poor countries.

Substantial international assistance (both technical and financial) may be needed in the future to facilitate adjustment to the changing macro-economic conditions at a time when: the favourable position in development aid is being eroded, preferential arrangements may be dismantled, opportunities for work abroad are becoming scarcer, local market sizes are limited, the resource range is becoming narrower, and increasing environmental problems and proneness to disasters jeopardize production.

Sustainable Management

Integrated Coastal Area Management

The continental shelves and coastal ecosystems of small island developing States are of major economic significance for settlement, subsistence and commercial agriculture, fisheries and tourism. Demands on coastal resources are endangering the long-term supply of these resources: large parts of coastal areas are being polluted by local or upland resources, fisheries over-exploited and fish habitats degraded, coastal belt mangroves cut, wetlands drained, coral reefs destroyed, fresh water aquifers are subject to depletion and salinity, bio-diversity conservation is threatened, and opportunities for aesthetic enjoyment of scenery reduced. Of particular concern to low-lying small islands is damage to natural systems such as coral reefs, dune lands, and coastal mangrove belts, which significantly reduces the protection they provide against natural disasters of marine origin (*i.e.* tsunamis, tidal surges and wave erosion).

The disruption of small islands societies' traditions sometimes by policy action or strong external factors through accelerated development, demographic growth, high rates of urbanization and often strong emigration fluxes are weakening traditional mechanisms of control, thus rendering coastal resources increasingly vulnerable to over-exploitation. Added to these sources of disruption are new problems such as more frequent natural disasters and rapid environmental changes which result in the degradation of terrestrial and marine habitats in coastal areas.

Increasingly in recent years, it is being understood that the search for solutions to conflicting demands in coastal areas must involve the various users and administrations concerned. With respect to land tenure, historically, small islands featured hierarchies of land ownership rights from which social status was derived. Environmental regulations limiting land use and controlling access to natural resources have been challenged by landowners who interpret any attempt to limit land rights as a challenge to their social status. The involvement and support of traditional leaders in introducing sustainable development practices is also necessary.

By the mid-1980s, the weakness of conventional sectoral planning in reaching sustainable solutions to conflicting demands for coastal resources was apparent. In response, multi-sectoral, integrated approaches to coastal area management began to be developed, in developed countries primarily, but also in some developing ones. UNCED gave support to this process by stressing the importance for small island developing States to develop national policies and management capabilities for multi-sectoral use of coastal areas.

Conservation and sustainable management of resources of coastal areas is known as integrated coastal area management (ICAM). This offers the means to balance the competing demands of different users of the same resources and to manage the resources sustainably.

Most archipelagic developing States have not established systems for the evaluation of their resources, nor the institutional and technical capacities to formulate and implement sound multiple-use strategies. A characteristic common to all small island developing States is the small number of skilled and experienced people available in most fields of activity. Finding people with the necessary experience to staff a government institution responsible for coordinating multiple use of resources, and taking an appropriate role in pro-active management will tax many island administrations.

Coordinating between sectors will be easier if line ministries and users of natural resources recognise their respective roles in the ICAM process and the impact of their activities on resource users in the sector itself and in other sectors. They must also have the capacity to monitor and assess the effect of the activities of other sectors on them.

Though of limited extents, the ecologies of most small islands are complex and varied. Detailed analysis of agro-ccological conditions is therefore required to provide an adequate form of integration and define improved multiple use of the natural resources.

The agriculture, forestry and fisheries sub-sectors are the chief users of resources in the coastal areas of small island developing States. To assist line ministries in their tasks of designing and implementing ICAM, guidelines such as those being prepared by FAO on the integration of agriculture, forestry and fisheries in integrated coastal area management will be most helpful-Guidelines can also be of assistance for planners and resource users at sub-sectoral level. FAO has also prepared guidelines on topics such as the environmental management of coastal aquaculture, the mapping and inventory of mangroves, and coastal rice culture.

FAO considers it important that appropriate emphasis be placed on specific training in integrated management for line agency staff. In this connection, it is collaborating with other UN agencies in the plan for human resources development and capacity-building for the planning and management of coastal and marine areas.

FAO has pioneered the concept of integrated coastal fisheries management (ICFM) as part of ICAM. The basis of this concept is twofold. First, the sea has long been seen as a bottomless sink into which flow pollutants produced on land which degrade marine habitats. Second, due to the open access nature of most fisheries, in many developing countries, small-scale fisheries become the employer of last resort when there is a surplus of unskilled labour in land-based sectors. This results in decreasing yields from already over-stressed fisheries. The open access nature of the resource also facilitates entry of industrial vessels into artisanal fisheries, contributing to overfishing and often resulting in social conflict. In many cases, it is in the interest of the fishery sector, far more than any other, to take a proactive role in initiating integrated management of natural resource use.

"Integrated Coastal Fisheries Management in Trinidad and Tobago"

Within a broader framework of ICAM, the UNDP/FAO project (1993-94) addresses the inter-relationship between fisheries management and other sectors in coastal areas of the Gulf of Paria including: urban development and housing, reclamation of wetlands for agriculture, run-off of agricultural inputs, depletion of natural forests in upland areas, hydrocarbon exploration and processing, and shipping. Data gathering and research for a better scientific understanding of coastal resources, socio-economic and cultural aspects of fishing communities and other users of the coastal and marine areas, greater awareness about the management needs of the fisheries and the coastal area, and the joint action of a number of government (local and national) agencies and users are the main components of this pilot project.

Land and Biotic Resources

Land-use Choices

The small size of many island developing States makes competition for land more severe than in larger countries. Large commercial plantations were frequently established under the colonial system, and continue today in the more fertile plains of island countries, forcing many small farmers to resort to the poorer soils in hilly regions. This, combined with poor land use practices, has resulted in deforestation and soil erosion.

Due to the peculiar geological characteristics of many islands, water resources are scarce, often limited to thin lenses of freshwater floating on sea water, recharged by rainfall. Poor management of watershed catchment areas and of water flowing through agro-ecosystems, particularly irrigation supplies, can lead to water scarcity and pollution of downstream supplies. Population pressure, expanding tourism, and reduced precipitations of recent decades, have resulted in the over-exploitation of groundwater and seawater intrusion, with subsequent further deterioration of water quality and quantity.

Land use planning is a prerequisite for long-term sustainability, since without it, the economic and social fabric of society as a whole will be endangered. Advice on area-specific soil and water conservation, and management practices will be needed for farmers, graziers and others. Such advice will only be adopted if it is perceived as being capable of bringing rapid economic returns to the users without threatening their fragile survival strategies. Success will also depend on guaranteeing long-term rights to land and water resources.

"Land Use Planning, Management and Information System in Grenada"

Grenada is highly dependent on agriculture, deriving 90 percent of its export earnings from this source, though the sector has stagnated over the past three years due to a combination of low export prices and declining productivity of the traditional export crops. The agricultural development strategy of the Government is to improve the productivity of traditional export crops, while diversifying into nontraditional crops for export and for agro-industrial integration into the expanding tourist sector. Emphasis will also be placed on increased domestic food production. The strategy involves bringing idle private and public lands back into production, and strengthening private and public sector institutions. Topography is a key factor in implementing the above strategy.

On the islands of Grenada and Carriacou, approximately 77 percent and 54 percent respectively of the land area has slopes exceeding 20 degrees. Only 33 percent of the soils are deep, well drained, and free of cultivation constraints. The rest, apart from excessive slopes, are shallow or rocky, or have seasonal drainage problems. A favourable feature of Grenadine geography is a well-defined system of watersheds governed by the topographic position of the rivers and streams.

The system of watersheds offers the opportunity for the establishment of rational land use consistent with the characteristics and conditions of each watershed, and with the need for soil and water conservation. The main objectives of this newly launched project (1993-94) are to create the technical basis to support rational development and conservation of the country's natural resources, and to develop an effective forestry management policy, and forest area conservation and development programme for the country.

The project will provide specialised technical assistance and practical hands-on training to local technical staff in the establishment and development of computerized natural resources information systems. The project is designed to provide a rapid response to an immediate need affecting the country's food production and the degradation of its natural resources. It is expected to contribute to establishing a sound basis for natural resource planning and policy formulation at both national and watershed levels. More specifically, the following needs to be emphasised:

- sound cropping practices and cropping systems which maintain soil fertility and productivity for sustainable crop yields as well as practices such as soil testing and analysis to determine fertilizer recommendations on a soil-and crop-specific basis. Land evaluation/land use planning are pre-requisites for sound land use and crop diversification. Development of land information systems will make a vital contribution to improved resource management;
- suitable water harvesting methods, both for domestic and agricultural use (supplementary irrigation), aiming at significantly increasing the groundwater recharge as well as collection and storage of rainwater from ground catchments;
- use of biomass residues (*e.g.* bagasse from the sugarcane industry), as a source of plant nutrients, will reduce use of mineral fertilizers in favour of nutrient sources available within the farm. This would contribute to environmental protection and decrease pollution of soils and groundwater.

Land conditions are similar in all coral island countries, and, apart from those affected by human action, soils are homogeneous. The main restrictions on land use are urban encroachment, tourism, established tree crops, especially coconut, or the existence of improved soils devoted to market gardening. Where choices for land use are still available in these islands, integrated coastal area management should be applied in a comprehensive programme of land use planning.

In small low-lying coastal areas, land well suited for agriculture is under severe pressure from expansion of housing, industry, and tourism. Under the market system, with private land ownership, the immediate economic returns from such uses of land always override the longer-term benefits to be derived from sustainable but low-return agriculture.

Economic pressures thus make it difficult for the rural sector to opt for the sustainable use of the resources entrusted to its care, such as:

- light use of coastal mangrove forests, while maintaining the additional functions of nursery areas for marine species, and protection against major damage to adjacent land from cyclones or tidal waves;
- brackish-water fish ponds in humid climates or salt pans in and or semi-arid conditions;
- cultivation of wetland rice or appropriate dryland crops, irrigated or rainfed, with flood protection and water control (mainly drainage facilities);
- cultivation of annual crops, adaptable to changing conditions, or perennial crops, entailing long-term decisions and larger investment; and
- conservation of local germplasm offered by traditional farming systems.

Hilly or mountainous areas face a different, but no less difficult, set of choices. Depending on climate, geology, landform and soils, the land may have high or low potential for nature conservation, forestry, grazing, production of perennial or long-term crops or cultivation of annual crops. Conditions can vary greatly over short distances, and the optimal response to a country's needs and objectives may well be a mosaic of different land uses and protected areas.

In small island developing States, the relationship between forests and trees, and other types of land use and economic activities, is critical. Forestry should be systematically represented in national and district planning commissions, land use zoning activities, town planning, energy and infrastructure strategies, as well as in planning for generation of income, employment, and food security. Forests and trees and the products and services they provide must be validated and valorized in view of their role in the protection of water resources and of the environment in general.

"Tropical Forests Action Programme"

Forest resources contribute to environmental protection and food security. Rural people exercise important customary rights over forest products, which have traditionally been used sustainably. Current forestry problems originate, among others, from official policies which are not supportive of sustainable management and utilization of forest and tree resources, The importance of forest resources has motivated many Caribbean island countries to develop appropriate inter-island forestry and environmental networks to promote forestry research, as well as carrying out Tropical Forests Action Programmes (TFAP) in order to implement the sustainable management of forest resources. The TFAF planning phase in Antigua and Barbuda, Barbados, Dominica, Grenada, Montserrat, St. Kitts and Nevis, St. Lucia, St. Vincent and the Grenadines, and Trinidad and Tobago is concluded and the implementation started in November 1993. This multinational exercise defined regional issues and produced key documents, including a survey of forest management and silvicultural systems in the region.

In the Pacific, Fiji is already implementing a TFAP-framed national forestry action plan, Vanuatu and Solomon Islands are starting the planning phase, and others (Kiribati, Micronesia, Tonga, Tuvalu, and Western Samoa) have expressed interest in launching a TFAP process. A TFAP Training Workshop for this region is scheduled for May 1994. In Africa, Cape Verde Islands are in active preparation of their National Forestry Action Plan, and a recently held national Forestry Congress has highlighted the overall progress achieved and original aspects of forestry in the country. Legislation and planning exercises are under way in the Comoros Islands.

Watershed Management

Urgent implementation of watershed management and rehabilitation programmes in most small island countries is needed for sustainable development. On most small islands, distances between highlands and coastal areas tend to be short. Under such conditions, the role of forest ecosystems as regulators of water supplies for the island crop environment and soil formation and protection is especially critical. By extension, the sustainability of agricultural productivity and potential greatly depends on the regulatory functions of forests.

Protecting forests and trees is essential for warding off environmental degradation and rural poverty. The multiple goods and services available from forests offer many possibilities for off-farm employment and for economic diversification, with low capital input, for example through the integration of plantations with agricultural crops, in soil conservation and watershed management, as recreational areas, and as a source of subsistence.

In spite of the significance of forests and tree-based resources, present trends are not encouraging; forest resources continue to be poorly managed and not used rationally in many island countries. Deforestation occurs due to clearing for agriculture in areas not suitable for this purpose, culling for fuelwood, and urban and infrastructure development. Population growth and economic reforms pose new and increasing demands on the forestry sector. The result is accelerated loss of forest cover and of its associated benefits.

"Watershed Management and Conservation Education in Western Samoa"

This UNDP/FAO project (1992-95) looks at the inter-relation between the management of upstream and downstream areas within a given watershed. This has been shown through the implementation of soil conservation and watershed management measures such as conservation plantation, agroforestry practices, plantation of forest and fruit trees by farmers, conservation farming systems, extension and education programmes for different target groups.

A monitoring system has also been developed to increase understanding of the impact of upstream management on water quantity and quality and other downstream values.

The destruction of vegetative cover by frequent tropical cyclones has led to accelerated soil erosion and landslides, with a direct impact on the watersheds, agricultural production, water supplies, hydropower generation, lagoons and reefs. The rehabilitation of upstream areas and the participation of the local population in conservation of downstream areas an integral parts of the project. This will facilitate the conservation of flora and fauna and improve the socio-economic condition: of watershed dwellers.

Greening Activities

Considering the limited land area and the islands' comparative fragility, the role of vegetation takes on particular importance. Some islands have no land suitable for agricultural activities and are confined to producing vegetables and growing fruit trees. The cover of low-lying islands is composed of mangroves and other plant associations, as well as exotic species. Demands for boats, construction material, rural housing, resort development and fuelwood have resulted in depletion of tree cover. This has affected the dynamic influence of ocean, coral reefs, land formations, and vegetation on each other.

Drought and desertification in some small island developing States renders afforestation difficult. In order not to interfere with endemic species and corresponding life systems by introducing exotic plants, it is preferable to use endemic species (bringing them in from nearby islands if local vegetation has been degraded) when undertaking forestry plantation and afforestation. Current water harvesting techniques, including soil manipulation devices aimed at concentrating rain water to better benefit planted seedlings, make plantation efforts in dry environments less uncertain than hitherto.

Further research is necessary for:

* improved knowledge of the original indigenous plant species and formations;
* selection and use of local species suitable for incorporation in soil and water conservation, range regeneration and fodder production, and fuelwood supply; and
* screening and eventual introduction of suitable non-indigenous species tried under similar conditions for use in all the activities mentioned above.

Greening activities should also include the identification, protection and proper management of natural formations.

"Development of Forest Resources in Cape Verde"

Within the national development framework, this project (1984-93) offered institutional support to the Government for the establishment and consolidation of an efficient forestry administration and the creation and management of enhanced forests and rangelands to promote rural development. These activities will contribute to combating, desertification, rehabilitate the environment through the reconstitution and conservation of the vegetative cover, establish integrated production system, contribute to the provision of wood for multiple uses (energy, construction, etc.), enhance socioeconomic welfare through the sustainable exploitation of forest resources, and transfer management responsibilities to a competent administration working within an adequate institutional framework. This project, completed in December 1993, should 'facilitate, *inter alia,* the finalization of the National Forest Action Plan, leading to better coordination in the planning and implementation of sector policies while promoting the participation of rural people in the preparation and management of the action plan.

Protected Areas and Biological Diversity

Island bio-geography has led to a high endemism in terrestrial species, especially in larger islands (*e.g.* over 1,000 indigenous species in Fiji). All small island developing States are particularly vulnerable to the impact of introduced species and deleterious human action.

The physical characteristics, the biotic formations, and socio-economic situation of small island developing States determine the conditions and role of forestry in sustainable development. The implications of their small size, their isolation, and the nature of their forest resources define the available models and opportunities for forest development, and suggest the necessary precautions to be taken in any manipulation of plant and animal populations. The increasingly negative impacts on local economies and on the quality of the environment are already visible in many cases through shortage of water supply, soil erosion, increased sedimentation of rivers, estuaries and coastal areas, and coastal erosion. Deforestation and forest degradation have led to the extinction of many animal and plant species, causing irreversible loss of genetic resources and ecosystems. The establishment of protected areas, such as forest reserves, national parks and wildlife sanctuaries, supported by botanical gardens, herbaria, etc., are essential for conservation of biological diversity in these countries. Those areas can become the basis for eco-tourism.

Marine ecosystems are prone to destruction of habitats and bio-diversity. There are several causes of this, including: destruction of coral reefs (both deliberately by fishermen or tourists and indirectly by pollution, sedimentation, and land reclamation for building sites); conversion of mangroves and wetlands (resulting in loss of important nursery areas); the use of long driftnets (mostly impacting marine mammals, turtles, and non-targeted fish) by a growing number of foreign fishing vessels; and more generally over-fishing. Lack of monitoring makes it difficult to quantify the damage of such activities on marine life. Surveys are needed to identify representative habitats and ecosystems for the establishment of marine reserves and parks.

Plant Genetic Resources

Plant genetic resources are both the building blocks of all living matter, and the raw material for the fast growing plant breeding and biotechnology industries. According to recent studies, any region of the world is dependent on genetic material which originated in other regions for over 50 percent of its basic food production; and for small island developing States, it may be close to 100 percent.

The conservation *(in situ* and *ex situ)* as well as utilization and management of plant genetic resources require both financial and human resources that may not always be available in small island developing States. Cooperation between those States is therefore desirable in this area to safeguard these resources for future use. It is of particular interest for small countries to have access to genetic resources from other countries in the same agro-ecological zone. Means are needed for conservation and evaluation of existing basic plant genetic resources, and for capacity-building for evaluation of crop cultivars for practical farming. The latter because sustainable farming systems need access to crop cultivars with an optimal production potential, good resistance and quality characteristics, and adapted to local environments.

Much of the plant genetic diversity still in active use in developing countries is maintained by small farmer-breeders as part of their livelihood and personal use. FAO has developed, within its International Undertaking on Plant Genetic Resources, the concept of Farmers Rights which recognizes the contribution of generations of rural people to domestication of wild species and their subsequent improvement through on-farm breeding. Farmers have thus conserved germplasm and made it available to societies and continue to do so. While it would be utopic for them to claim individual compensation, the concept form the basis of formal recognition and reward system to encourage and enhance the continued role of farmers and rural communities in the conservation and use of plant genetic resources. This aims at reconciling the view of the "technology rich" and the "gene rich" countries in order to ensure the availability of plant genetic resources within an equitable system.

"Improved Seed Production in the Caribbean"

Production of good quality seed is very limited in the Caribbean. With the exception of sugarcane, there is practically no germplasm information or exchange of species and variety performance among countries. In general, countries lack the infrastructure and technical skills to produce the required seed. Selected seeds (mainly vegetables) are imported, in most cases without any quality control or adaptation trials this results in frequent unsatisfactory germination and low yields. This project (1992-94) is developing a regional seed technology training programme and an international information exchange network on plant genetic resources availability, variety descriptions and variety trails. Field and laboratory standards are established for seed quality control for both sexual seed and vegetative planting materials, to facilitate active germplasm and variety exchange programmes among countries (Antigua and Barbuda, Bahamas, Barbados, Belize, Dominica, Grenada, Guyana, Jamaica, Montserrat, St. Kitts and Nevis, St. Lucia, St. Vincent, Surinam, and Trinidad and Tobago). Specific seed activities will be supported in selected countries. It is intended in a second phase to promote active germplasm exchange, seed production, and further development of seed technology.

The concept of Farmers Rights also provides the counterbalance to "formal" Intellectual Property Rights and Patents which compensate for the latest innovations (technologies and/or others) without giving enough consideration to the fact that in many cases, these innovations are only the last step of accumulative knowledge and innovations carried out through many generations over millennia in different parts of the world.

Small farmer-breeders are generally beyond the reach of the seed distribution infrastructure and maintain actively considerable genetic diversity. The Convention on Biological Diversity has left-out two outstanding issues: access to ex-situ collections and farmers rights. In order to address these issues, FAO has initiated a countries' negotiating process to revise the International Undertaking on Plant Genetic Resources which would be presented to the Conference of the Parties on the Convention on Biological Diversity as a possible protocol.

During the late 1980s, FAO assisted the countries of the South Pacific region for collection, conservation, characterization and documentation of root and tuber crop germplasm. Currently most of the collections are held in the Dodo Creek Research Station, Honiara, Solomon Islands. In the coming biennium, FAO proposes to survey the Pacific region to evaluate the status of existing *ex situ* collections and to propose action to address any related problems and strengthen national capabilities for conservation and sustainable utilization of existing germplasm collections.

Managing Risk due to Natural Hazards

Cyclones inevitably affect any programme dealing with sustainable agriculture and rural development. Early-warning systems and disaster preparedness, benefiting from the collaboration of several international organizations to minimize the impact of cyclones, are already under operation in many countries.

The effects of hurricanes and cyclones on forests and forestry plantations are a major constraint on sustainable forestry development in the Pacific, Indian Ocean and Caribbean islands.

Research and development into forestry and natural resource conservation and management are needed to mitigate the effects of strong winds and cyclones on these countries, in particular:

* salvage logging after windblow, including inventory, logging techniques and the storage of large quantities of logs before processing;
* selection of species and tree breeding through hybrids or other improved material of wind-firm trees;
* tree breeding strategies where trials or seed trees are liable to be destroyed; and
* management techniques especially related to reduction of damage through initial spacing, pruning and thinning.

There is thus a strong need to develop technology-transfer programmes to promote environmentally sound forest operations which take account of the special environmental risks associated with small island developing States.

"High Wind-resistant Trees"

In 1991, Cyclone Val destroyed about 60 percent of the vegetative cover in the Vaisigano watershed area in Samoa. The result was accelerated soil erosion and landslides, and increased surface run-off. Agricultural production, water supplies for consumption and hydropower generation, lagoons and reefs were all affected. As a quick rehabilitation-measure, the area has been planted with *Albizzia* species at the spacing of 10 x 5 metres to allow the regeneration of natural vegetative cover.

This tree species was selected because of its resistance to high winds (even when badly damaged by cyclones. *Albizzia* revives), low-labour requirements (being a fast growing species, no maintenance is required after 1.5 years), capacity to improve soil fertility and ability for self-regeneration within 4-5 years (thus providing quick cover). In Fiji, the pine *(Pinus Caribaea)* forests are liable to exposure to storm or hurricane force winds at least twice during a pine rotation. Wind damage to the forests is often severe and occasionally devastating. Research over several years has aimed at improving crop stability in plantations. Achievements include the establishment of clonal seed orchards introducing a group of more windfirm coastal provenances of pines selected from survivors of hurricane force winds. Species trials have shown that several hardwood species have greater vigour and stability than *P. caribaea* and may be valuable as shelterbelts on very exposed sites.

Apart from the direct impact of cyclones on crop production, the major risk to agriculture is salt-water intrusion on both soil and freshwater resources. Strong storms can result in complete coverage of low-lying islands by waves, preventing subsequent growth of crops for many months due to residual salt. Most importantly, salt water contaminates the limited freshwater lense underlying atoll islands which is the only source of water for both drinking and agricultural purposes.

The gravity of salt-water intrusion into the freshwater lense would he exacerbated if the predicted sea-level rise materialises: lateral leakages would increase, lenses would become thinner, salt-water would reach water-pumps and wells, and the roots of tree crops such as coconut palms and breadfruit. Agricultural insurance is available for weather damage and the basic requirements for this are well documented in FAO. The Organization is in a good position to provide advice on the management of the attendant risks, particularly adverse weather events, such as cyclones.

"Banana Insurance in the Windward Islands"

Bananas are the principal export crop of the Windward Islands (St. Lucia, Dominica, Grenada, St. Vincent). The crop is vulnerable to excessive winds. Individual growers, (totalling some 25,000) who export through the group marketing scheme, WINBAN, are compulsorily covered under an insurance scheme run essentially on commercial lines, and embracing all four island nations of the Windward Group. The premium rate charged is actuarially determined and is currently approximately 13 percent of the sum insured. The scheme is entirely funded by the premium income. Given the catastrophic nature of hurricanes and the possibility of massive losses, re-insurance is arranged by WINCROP on the international market. Loss assessment is carried out by WINCROP staff, assisted by on-call assessors, as required. Records for the entire operation are computerised, which means that loss assessors are provided with a computer printout giving details of each banana operation-area, past production, past loss assessment-a key to enabling this scheme to run efficiently, without subsidy.

Rural Energy Resources

People living on small islands frequently face difficulties in maintaining a basic standard of living; isolation, high transportation costs, limited markets for services and goods and the lack of indigenous resources are factors that affect every aspect of life. Most marketed goods are expensive or difficult to obtain, not least a reliable energy supply or domestic fuel.

In most islands, the better-off people enjoy the range of household or industrial facilities common to cities and towns everywhere. Urban communities on most islands depend upon a municipal electricity supply. Power is normally generated centrally from oil or gas delivered to the island and, increasingly, from indigenous resources such as agricultural residues.

Renewable sources of energy are sometimes available, including geothermal sources, wind generators and hydro-electric turbines, but they are relatively limited compared to conventional generation (usually the internal combustion engine), mainly because the costs of exploitation and use normally makes them uneconomical compared to imported fuels.

Notwithstanding the dependency that this implies, and the need to obtain foreign currency with which to purchase oil or gas, there are significant advantages to utilizing commercial sources of power and equipment. Liquid or gas fuels are versatile and compact, and the range of services and equipment available with which to use them is extensive, at least for urban dwellers.

Providing electricity to rural communities is, however, costly and this may offer opportunities for developing renewable sources of power. Currently, the range of equipment and services for using alternative energy resources is limited, often expensive and sometimes difficult to obtain. Moreover, much of the equipment is site-specific, for example, wind powered equipment.

Government policies can encourage the introduction of renewable sources of energy. This has been the case with wind-generators in Denmark during the past 20 years. The use of renewable sources of energy, though sometimes expensive in terms of direct costs, can be more easy to justify on environmental grounds.

Agricultural and forest biomass is the oldest form of fuel and remains the most common one for rural and urban households the world over, whether for heating and cooking or for cottage industries. Fuelwood sometimes constitutes an important source of income for those living around forest areas. For densely populated islands, exploitation of wood for fuel is no longer a viable option and dependence on imported fuels has generally been on the increase.

Wood for energy can be grown in plantations or woodlots, or derived from sustained management of forest woodland, agroforestry or trees outside the forest. This constitutes an important potential source of fuel compatible with conservation of the environment and is complementary to other important functions of forests and trees and associated housing areas. Bagasse as a feed for ruminants is well-documented and sometimes widely used. Household cooking stoves fuelled by rice husks have been developed and are popular in South East Asia, but rarely seen in other rice-growing areas. In many small island developing States, animal traction has an extremely important role to play as a source of energy. In the Dominican Republic, La Romana Sugar Estate, one of the biggest sugar factories in the world, has kept up to 18,000 oxen for transporting its sugarcane to the factory. Another increasing contribution of animals to energy production comes from biogas. New simple biodigesters made of PVC plastic sheets are now used in various tropical countries to produce cheap energy for use in kitchens.

The role of FAO, and the other international agencies, in the field of energy resources for island developing States depends in the first placc on requests from governments for assistance. Such requests are few, perhaps due to the aggressive policies of the commercial power generation industry, particularly for petroleum. Renewable power generation industries also exist, but they are smaller and their capacity to influence policy is limited.

There is scope for providing information, contacts, companies, case-studies, etc. to enable people to compare and contrast experiences, and to obtain technical and commercial information. FAO can also assist with the preparation and execution of projects which focus upon energy supplies. These can cover a wide variety of activities, from fuelwood plantations to solar-powered telecommunications or cold-storage for food or medicines.

Macro-planning at national level, to assist governments (and the private sector), to determine longer-term strategic energy supplies is well within FAO's technical resources. However, so long as price is the sole criterion in determining energy supplies and environmental costs are not included in price calculations, alternative, renewable resources are unlikely to gain a larger share of the energy market.

"Bio-energy from Agricultural Residues"

A large number of small island developing States, mainly in the Caribbean and the Pacific, are major sugar producers and this agro-industrial activity remain their main source of income. Sugar mills are large consumers of energy and have long burnt sugarcane bagasse to generate the electricity and heat required in the production process.

Depending on the efficiency of the system, normally directly related to the age of the plant, these mills can be self-sufficient in energy terms, and can even produce extra power to feed to the electricity grid.

The amount of bagasse produced usually far surpasses the requirements of the mills, and in many cases, the disposal of bagasse is a major environmental problem. Many island States have embarked on a process of modernizing their sugar mills in order to improve their economic operation. Among the changes being introduced are better combustion and larger power generation systems.

The potential of these bioenergy resources is considerable. Production of ethanol as fuel, of animal feed and of fibre are among the potential uses and this could open new economic and employment opportunities to small island developing States. Nevertheless, technical and legislative measures need to be developed and implemented to overcome barriers to the full use of the energy potential of these biomass resources. In many countries, legislation still does not permit power generation and especially power feeding into the national grid by institutions outside the traditional power sector. Environmental policies will undoubtedly point towards the promotion of such schemes.

With moderate inputs and suitable management, existing varieties of sugarcane produce reasonable yields in marginal lands opening up new energy and environmental opportunities. FAO organized in 1993 an Expert Consultation on Biofuels for Development and plans to expand its activities in this: field through global and national studies. Biomass resources can contribute to raising incomes in the agricultural sector, and to the achievement of stable and self-sustainable energy systems.

Edibility of Ornamental Tree Fruit

Ornamental trees, such as crabapples and purple-leaved plums, are planted in the home landscape for their flowers or colourful foliage. While chiefly ornamental, these trees also produce fruit resembling those on fruit trees. As the fruits mature, questions concerning their edibility often arise. Fruit from ornamentals, such as crabapples and purple-leaved plums, are edible. However, the eating quality of some ornamental fruit is rather poor. These trees were selected for their ornamental characteristics rather than the eating quality of their fruit.

The crabapple (*Malus spp.*) is an example of a common ornamental that also produces edible fruit. Apples and crabapples are differentiated strictly on the size of their fruit. Crabapples are defined as those varieties with fruit 2 inches or less in diameter. Those with larger fruit are apples. At maturity, crabapple fruits vary from yellow to orange to bright red. The fruit of some crabapple varieties colour and ripen in August, others mature in the fall. (The colourful, persistent fruit of many of the newer crabapple varieties are actually an important ornamental characteristic.) While all crabapple fruit can be used in making jellies and preserves, large-fruited varieties, such as 'Whitney' and 'Chestnut,' are the best.

Purple-leaved plums (*Prunus cerasifera*), such as 'Newport,' 'Thundercloud,' and 'Mount Saint Helens' are planted in the home landscape for their white or pink flowers and purple foliage. In most years, ornamental plums produce little or no fruit as they bloom so early that pollination is generally poor. Occasionally, however, they produce a good crop of fruit. At maturity, the edible fruits are approximately 1 inch across and dark purple in colour.

The new growth of 'Canada Red' and 'Schubert' chokecherry (*Prunus virginiana*) emerges green, but quickly turns to a maroon red. The small trees produce clusters of white flowers in spring followed by round, 1/3-inch diameter fruit. At maturity, the fruit are dark purple to black. Though bitter, the mature fruit can be used for jams, jellies, and juices.

'Autumn Blaze,' 'Chanticleer,' and 'Redspire' are ornamental pears (*Pyrus calleryana*) grown for their glossy dark green foliage and small white flowers in early spring. Ornamental pears generally produce only small numbers of fruit. The inedible fruit are roundish, 1/3-inch across, and tan to brown in colour. A word of caution. If you are uncertain of the identity of a tree or shrub, don't eat the fruit. It's better to be safe than sorry.

Incredible, Edible, Ornamental Sweet Peppers

Think of a vegetable whose fruits can be green, oranges, red, gold, lime or even purple. Combine this broad colour spectrum with crunchy texture and mellow sweet taste, endless culinary uses, ease of growth and compact plant habit, and you've got a versatile ornamental vegetable — the sweet pepper.

From the classic bell type (referred to as a "mango" in some areas) to its relatives the 'Cubanelle,' pimiento, 'Sweet Banana', 'Sweet Cherry' and sweet wax types, sweet peppers offer a palette of culinary and visual pizzazz.

Sweet peppers are certainly one of the premier edible ornamentals. In recent years they've emerged from the vegetable patch and are now found lending their decorative, festive touch to flower borders, patio gardens, entrance yards, porch boxes, containers, and all types of locations calling for productive, colourful plants.

Even the most common pepper type, a green bell such as 'California Wonder,' provides a visual delight. Much to the surprise of many novice gardeners, the normally shiny green fruits develop further to brilliant red when left on the plant to full maturity. A check at the grocery produce counter shows that not only the colour changes, but the fruits' value goes up, often by a factor of two to three times. Red bells are also far sweeter and more nutritious than green ones, and their mellow taste makes the wait for maturity worthwhile.

Pickling Favourites

Look to the smaller-fruited sweet peppers for more variety and sheer numbers of fruits. Types such as 'Sweet Cherry' and 'Hungarian Sweet Wax' consistently produce three dozen or more fruits per plant. The golf ball size fruits of 'Sweet Cherry' are shiny green at first and finally turn brilliant red, making them perfect for relish trays or salads. Their shape and small size make them favourites for pickling whole. A single 'Sweet Cherry' plant is tailor-made for growing in a three-gallon container, or for combining in a larger container with other vegetables and flowers. The sturdy, bushy plant seldom grows taller than 20 inches and sparkles with its crop of bright fruits.

'Hungarian Sweet Wax' is another favourite for pickling. Its 5- to 6-inch long tapered fruits come in a mix of yellow, orange and red, depending on degree of maturity. They pack easily into quartsize canning jars, making an attractive display.

Watch seed catalog descriptions and bedding plant labels carefully because you don't want to confuse 'Hungarian Sweet Wax' with its look-alike, but fiery hot cousin usually sold as 'Hungarian Wax' or 'Hungarian Yellow Wax.' Most seed catalogs list hot varieties separately, but at bedding plant outlets, labels are sometimes inadvertently switched.

'Sweet Banana' is a prolific producer of fruits with size and shape similar to those of 'Hungarian Sweet Wax.' Colours range from lime-green through yellow, orange and finally red. A pickling favourite, 'Sweet Banana' is as good as nearly any pepper for fresh use, and is widely used for frying. 'Top Banana,' a new hybrid introduction this spring, is described as "outclassing open-pollinated varieties with prolific production and much better performance under stress." 'Top Banana' matures its first fruits in about 63 days from transplanting, making it a likely choice for short season areas.

Few varieties beat 'Cubanelle' when fried peppers are on the menu. The pendent, tapered fruits usually reach a length of 6 inches and measure 2 inches or slightly more at the shoulder. 'Cubanelle' is best harvested at the yellow-green stage, when the fruit walls are still firm and hold up well. Fruits turn red and become sweeter at full maturity, but lack substance at this stage.

In Northern Zones

Gardeners in short season areas can expect best production from early maturing varieties. Assist these carefully selected types with season extenders such as cloches or similar covers, and heat-trapping mulches. Utilize a planting area facing a south or west wall.

This may make the difference between success or failure in a marginal climate. Plants growing in containers have the extra dimension of portability, permitting you to move them to ideal locations for growth, and later to a spot where their colourful, edible fruits can be enjoyed.

As you plan this year's garden, keep the ornamental versatility of sweet peppers in mind. Give them a try in a few locations beyond the vegetable garden. You'll be rewarded with easy-to-maintain ornamental plants, and abundant harvests of attractive, nutritious and delicious fruits.

6

Scientific Weed Management Strategy

Weed

A weed in a general sense is a plant that is considered by the user of the term to be a nuisance, and normally applied to unwanted plants in human-made settings such as gardens, lawns or agricultural areas, but also in parks, woods and other natural areas. More specifically, the term is often used to describe native or non-active plants that grow and reproduce aggressively. Generally, a weed is a plant in an undesired place.

Weeds may be unwanted for a number of reasons: they might be unsightly, or crowd out or restrict light to more desirable plants or use limited nutrients from the soil. They can harbour and spread plant pathogens that infect and degrade the quality of crop or horticultural plants. Some weeds are a nuisance because they have thorns or prickles, some have chemicals that cause skin irritation or are hazardous if eaten, or have parts that come off and attach to fur or clothes.

The term *weed* in its general sense is a subjective one, without any classification value, since a "weed" is not a weed when growing where it belongs or is wanted. Indeed, a number of "weeds" have been used in gardens or other cultivated-plant settings. An example is the corncockle, *Agrostemma*, which was a common field weed exported from Europe along with wheat, but now sometimes grown as a garden plant.

Professor Richard C. Lewontin of Harvard University defines weeds as plants that create environmental conditions in which it cannot reproduce. He takes the example of pine trees that crowd out sunlight such that its own offspring cannot grow. Weeds continue to exist, because the environment is continually being disturbed to create open conditions for new generations, such as forest fires and human activity.

Distribution

Weedy plants generally share similar adaptations that give them advantages and allow them to proliferate in disturbed environments whose soil or natural vegetative cover has been damaged. Naturally occurring disturbed environments include dunes and other windswept areas with shifting soils, alluvial flood plains, river banks and deltas, and areas that are often burned.

Since human agricultural practices often mimic these natural environments where weedy species have evolved, weeds have adapted to grow and proliferate in human-disturbed areas such as agricultural fields, lawns, roadsides, and construction sites. The weedy nature of these species often gives them an advantage over more desirable crop species because they often grow quickly and reproduce quickly, have seeds that persist in the soil seed bank for many years, or have short lifespans with multiple generations in the same growing season. Perennial weeds often have underground stems that spread out under the soil surface or, like ground ivy (*Glechoma hederacea*), have creeping stems that root and spread out over the ground.

Some plants become dominant when introduced into new environments because they are freed from specialist consumers; in what is sometimes called the "natural enemies hypothesis," plants freed from these specialist consumers may increase their competitive ability. In locations were predation and mutual competitive relationships no longer exist, some plants are able to increase allocation of resources into growth or reproduction. The weediness of some species that are introduced into new environments can be caused by the introduction of new chemicals; sometimes called the "novel weapons hypothesis," these introduced allelopathyic chemicals, which indigenous plants are not yet adapted to, may limit the growth of established plants or the germination and growth of seeds and seedlings.

Relation to Humans

As long as humans have cultivated plants, weeds have been a problem. Weeds have even been mentioned in religious and literature texts like the following quotes from Genesis and a Shakespearean sonnet:

> *"Cursed is the ground because of you; through painful toil you will eat of it all the days of your life. It will produce thorns and thistles for you, and you will eat the plants of the field. By the sweat of your brow you will eat your food until you return to the ground,"*
>
> *"To thy fair flower add the rank smell of weeds: But why thy odour matcheth not thy show, The soil is this, that thou dost common grow."*

Weed seeds are often collected and transported with crops after the harvesting of grains. Many weed species have moved out of their natural geographic locations and have spread around the world with humans. Not all weeds have the same ability to damage crops and horticultural plants or cause harm to animals. Some have been classified as noxious weeds by governmental authorities because if left unchecked, they often dominate the environment where crop plants are to be grown or cause harm to livestock.

They are often foreign species mistakenly or accidentally imported into a region where there are few natural controls to limit their population and spread. Many weeds have ideal locations for growth and reproduction because of the large areas of open soil created by the conversion of land to field agriculture. Farming practices that produce unvegetated soils part of the year and human distribution of food crops mixed with seeds of weeds from other parts of the world have facilitated the colonization of vast new areas for many weedy species; humans are the vector of transport and the producer of disturbed environments, thus many weedy species have an ideal association with humans.

A number of weeds, such as the dandelion *Taraxacum*, are edible, and their leaves and roots may be used for food or herbal medicine. Burdock is common weed over much of the world, and is sometimes used to make soup and other medicine in East Asia. These so-called "beneficial weeds" may have other beneficial effects, such as drawing away the attacks of crop-destroying insects, but often are breeding grounds for insects and pathogens that attack other plants. Dandelions are one of several species which break up hardpan in overly cultivated fields, helping crops grow deeper root systems. Some modern species of domesticated flower actually originated as weeds in cultivated fields and have been bred by people into garden plants for their flowers or foliage.

Noxious Weeds

Most gardeners would say that any weed in their garden is a noxious weed. However, in Washington State, the term "noxious weed" has a specific legal meaning: a non-native plant designated as such by federal, state or county law as aggressive and/or injurious.

The State of Washington has enacted legislation which created a series of Weed Control Boards at the state, regional, and county levels. The purpose of these boards is to identify aggressive and/or injurious weeds within their jurisdictions, and to facilitate the control and removal of these weeds.

Each year the State Noxious Weed Board adopts a State Noxious Weed List. Regional and county weed boards use this list as the basis for creating their own noxious weed lists, based on the their local weed control priorities. Washington weed control laws assigns primary responsibility for noxious weed control to the landowner-whether it is private, state, or county lands.

The Impact of Noxious Weeds

Noxious weeds pose a significant threat to Washington agriculture and natural resources. Noxious weeds invade farm land and pastures, increasing the overall cost of farming and, in some cases, making land unsuitable for agriculture. Some of the noxious weeds that invade pastures are toxic to domestic animals. Noxious weeds in the natural environment may be thorny and injure wildlife, may be toxic to native wildlife species, and may crowd out native plants. The chemicals used to control noxious weeds may also damage the environment.

Aquatic noxious weeds can clog waterways, crowd out native plant species, prevent native fish from accessing breeding areas, damage or ruin commercial and recreational fishing, and make swimming dangerous.

The economic cost of controlling noxious weeds is borne by all of us, even if we live in an urban environment. Noxious weed control increases the cost of agricultural food and products. Tax money is required to remove noxious weeds from the edges of our roads. By law, noxious weeds must be controlled by parks, state forests, schools and cities, resulting in higher fees and taxes. Removing noxious weeds from our lakes and waterways, and then restoring them, is very expensive.

By controlling noxious weeds landowners protect life and health, improve habitat for fish and wildlife, maintain native plant communities, minimize the impacts to agriculture production, and save money.

Noxious Weed Classifications

The Washington State Noxious Weed Control Board adopts a State Noxious Weed List each year. This list categorizes weeds into three major classes (A, B and C) according to the seriousness of the threat they pose to the state or a region of the state:

- Class A weeds are non-native species with a limited distribution in Washington. Preventing new infestations and eradicating existing infestations is the highest priority. Control of these species is required by law.
- Class B weeds are non-native species that are presently limited to portions of the state. Class B species are designated for control in regions where they are not yet widespread. Preventing infestations in these areas is a high priority. In regions where a Class B species is already abundant, control is decided at the local level, with containment as the primary goal.
- Class C weeds are other non-native weeds found in Washington. Many of these species are widespread in the state. Long-term programs of suppression and control are a local option, depending upon local threats and the feasibility of control in local areas.

Noxious Weeds Of Greatest Local Concern

Nancy Ness, Coordinator of the Grays Harbour County Noxious Weed Control Board, provided a list of the Class B and C noxious weeds of greatest concern in our local area (All Class A noxious weeds are a high priority for control).

Weeds have been a part of civilization for as far back as recorded history. Many ancient documents speak of humans battling weeds in the crops they grew; where there has been agriculture, weeds have also been there. The spread of invasive noxious weeds on public lands in the West has been described by Steven A. Dewey of Utah State University as a "biological wildfire raging out of control." Wildfires and noxious weed infestation both may begin in a small area and spread slowly. The fire may spread quickly if embers are dispersed by the wind, creating "spot" fires which can grow and join each other and the main fire; weed seeds may be dispersed by wind, water or animals and create spot weed patches which grow and merge with the main infestation within a pasture, agricultural crop, recreation area, golf course, or backyard.

Wildfires and noxious weeds both dramatically change plant communities and may affect wildlife, livestock, crop production, soil erosion, water quality, or recreation; ecosystems extensively invaded by nonindigenous noxious weeds may be permanently altered.

Fire fighting is so similar to fighting weeds that modern wildfire management provides useful ideas on ways to improve noxious weed control programs. According to Dewey, wildfire management includes prevention (education and regulation), early detection and rapid response, containment and control, and site rehabilitation. Experience in developing and implementing effective public education programs in fire prevention could serve as a source of ideas to help increase the public's awareness, concern, and involvement in noxious weed prevention.

There are many definitions of a weed; the most commonly accepted definition is that a weed is a plant growing where it is not wanted. Almost any plant can be a weed if it is not wanted; however, plants that are commonly described as weeds can be desirable if they are growing on a hillside and controlling erosion.

Weeds are undesirable and considered to be pests just as insects and disease organisms are considered to be pests. They can be found everywhere that humans live or work, infesting lawns, clogging waterways, filling up ponds, invading flower beds, invading golf courses, and growing in pastures and forests. Weeds compete with crops for materials essential to growth, including fertilizer, water, sunlight and space.

Weeds may also serve as a host for insects and disease organisms, cause impurities in agricultural products, and be poisonous to livestock. Cows that graze in pastures where wild onion or wild garlic grow will produce milk that has an offensive odour and taste, which lowers the quality and value of the milk. Similarly, wool that contains burs or other parts of weed plants is less valuable because of the added expense of removal.

Tansy ragwort (Senecio *jacobaea)*, cocklebur *(Xanthium strumarium)*, nightshade (Solanum *nigrum)*, and bracken ferns *(Pteridium aquilinum)* are examples of plants that can be poisonous to livestock. Tansy ragwort grows in the northwestern United States and can be deadly? if a horse ingests this plant, its liver ceases to function normally death can occur in a short time. The removal of these weeds from hay fields and pastures is time-consuming and expensive.

Weeds cause problems for humans: many people suffer from allergic reactions to pollen produced from noxious weeds during certain times of the year, such as ragweed *(Ambrosia artemisiifolia)* and goldenrod *(Solidago spp.)*. In addition, weeds lower the aesthetic value of landscaped areas, including gardens, lawns, and golf courses.

The Classification of Weeds

Individual states determine which weeds are noxious, that is, the most difficult to control and create the biggest problem; obviously, a weed may be declared noxious in one state and not in another. Weeds are also classified according to their life span, annual, biennial, or perennial. The Federal Noxious Weed Act, Public Law 93, was enacted January 3, 1975 to establish a Federal program to control the spread of noxious weeds. The Secretary of Agriculture was given the authority to designate plants as noxious and the transport of all such weeds in interstate or foreign commerce was prohibited. The Secretary was also given authority to inspect, seize and destroy products, and to quarantine areas, if necessary to control, eradicate, or prevent or retard the spread of such weeds.

Characteristics of Weeds

Particular plants become troublesome pests because they grow and reproduce efficiently and out-compete agricultural crops; many weeds produce a large number of seeds, as many as 250,000 from a single plant. Many species have adapted efficient methods of seed dispersal: the milkweed *(Asclepiadaceae syriaca)*, thistle *(Cirsium odoratum)*, and dandelion *(Taraxacum officinale)* have long featherlike structures and may be carried by the wind for long distances; other seeds, such as cocklebur and beggar's lice *(Galium aparine)*, have projections on their seed coats that may catch in the hair of animals or the clothing of humans to be carried away.

Seeds from the fruit of weeds may stick to the feet or bodies of birds and be carried to new locations; other weed seeds may be ingested by animals and pass through their digestive tracts far from the parent plant; some weed seeds can be carried by farm machinery from one field to the next or travel on wagons or crates; and, some seeds may be carried by water in streams or dispersed onto fields from open canals during flood irrigation. Weed seeds generally remain viable over long periods of time, inactive in dormancy until the germination process is triggered by soil moisture, soil temperature, or light.

Weeds may reproduce by more than one method, both sexually and asexually. For example, the noxious weed yellow nutsedge *(Cyperus esculentus)* propagates by seed and rhizomes, underground stems that grow out from the plant, forming tubers and new shoots. In developing a method of weed control, a dual form of reproduction makes the task more difficult.

Imported Weeds

Of the fifteen of the most serious weeds in the United States, thirteen were imported. For example, Kudzu (Pueraria *lobata)* was introduced from Japan in the late 1800's to control erosion in large areas of the southern U.S. The kudzu vine grows quickly in thin soil and the dense foliage is a palatable, high protein forage for cattle; furthermore, since kudzu is a legume and dies back each year at frost, it adds organic matter and nitrogen to the soil.

Unfortunately, kudzu grows so quickly (up to one foot per day!) in a year's time, the aggressive vine can cover the ground or forest, blocking sunlight and killing trees; with its deep root system, it is difficult to control. Due to its prolific nature and lack of natural insect or disease controls, kudzu quickly became a pest in the southeastern United States; kudzu has the ability to out-compete and eliminate native plant species and upset the natural diversity of plant and animal communities.

Controlling Weeds

As methods of controlling weed pests improve, growing crops becomes more efficient and profitable. Billions of dollars have been invested in research and a variety of methods are available to control weeds, including mechanical, chemical, biological, cultural control.

People learned that simply pulling weeds by hand from around plants is time-consuming and disrupts the roots of desirable plants; when farmers learned to plant crops in long rows, hoes and plows were invented to cut off, uprooted, or cover up weeds. This method was used in the United States until the 1960s when chemical herbicides became readily available.

Mechanical cultivation to control weeds has four disadvantages:

1. The roots may suffer some collateral damage,
2. Equipment is expensive to purchase, operate and maintain,

3. Tillage can cause problems with erosion, and,
4. Cultivation results in soil moisture loss and may increase irrigation costs.

Chemical Weed Control

Centuries ago, farmers controlled weeds with salt; unfortunately, compounds such as salt are residual in the soil and prevent plant growth when concentrations are too high. With the invention of 2-4D around World War II, herbicides were developed rapidly and became widely used by the 1960s. Modern herbicides must meet rigid environmental standards set by the Environmental Protection Agency.

How do herbicides work ? The active ingredient of a specific herbicide may interfere with plant photosynthesis, inhibit plant enzyme production and cell growth, block cell division ("mitotic poisons") and plant growth, prevent the formation of carotenoids and other photosynthetic plant pigments, or disrupt auxins levels and plant growth.

Herbicide Selectivity? Nonselective herbicides are chemicals which kill all vegetation; selective herbicides kill only a particular plant type. Post-emergent chemicals are applied to growing crops absorbed after contact with the leaves, stems, or roots. Pre-emergent chemicals are applied to the soil surface before seeds germinate. Plants have different responses to these systemic herbicides; some weeds are unaffected by certain herbicides because they can produce an enzyme to neutralize the chemical poison. If crop plants can be genetic engineered to produce enzymes to neutralize herbicides, weeds can be more efficiently controlled. Research and development of new herbicides and herbicide-resistant crop varieties is expensive and requires years of extensive testing.

Biologic agents include animals, insects, and disease organisms. Goats may be kept in areas where sheep and cattle graze to eat the coarser, brushlike plants and make more room for grasses, geese have been used in cotton, mint, and strawberry fields to control grass, and fish and snails are used to control aquatic weeds in lakes and aquaculture ponds.

Importation and release of phytophagous (plant-eating) insects for control of weeds in Hawaii began in 1902 with introductions against lantana *(Lantana montevidensis),* a pest primarily of lowland pasture. During a span of more than 50 years, at least 27 species of lantana-feeding insects were released, 15 of which were successfully established: one species eats the flowers and seeds, another eats the leaves, yet another girdles the stem of the bush. In all, 71 species of phytophagous insects and one species of fungus had been purposely introduced and released up to 1985 to aid in the control of 21 weeds. The largest number of insect species introduced belonged to the Lepidoptera (31 species of butterflies and moths), followed by the Coleoptera (24 species of beetles), and Diptera (8 species of flies). To date, 43 of the insects and the fungus have become established.

Disease organisms can also be used to control weeds, including bacteria, viruses, and fungi. Researchers in Minnesota examined fifteen genera of fungi frequently isolated from purple loosestrife for their potential as biological control agents. The objective was to develop a mycoherbicide that would be applied alone or in combination with other agents, such as insects, to reduce purple loosestrife populations. The most virulent fungus tested was *Septoria lythrina.*

The use of biological control is relatively slow and the introduction of insects, animals, or diseases can be expensive. In the near future, genetic engineering may provide pathogens that attack only specific species of weeds; however, producers must be vigilant that agents introduced for biocontrol do not become pests themselves.

Cultural Methods of Weed Control

Equipment, such as cultivators and plows, can spread weeds from field to field, especially if the weed has rhizomes which cling to the bottom. The rotation of crops can disrupt weed life cycles and provide control options, especially if the weed is relatively easy to manage in an alternative crop. Black plastic mulch can be used to achieve almost total weed control.

The use of cultural methods of weed control in conjunction with mechanical methods, herbicides, and biological controls is called integrated pest management, or IPM. By using the best of all these methods, weed control can be effective and efficient, while protecting the environment. In the future, scientists will find even better ways to control weeds that are friendly to the environment.

Parameters for Weed-crop Competition

In most developing countries, agriculture employs more than three-quarters of the labour force and provides a major source of GDP with a share of 35-40 percent. A crucial policy issue is how to raise the income of resource-poor farmers without despoiling the natural resource base. A holistic approach based on Integrated Pest

Management (IPM), as well as on sound economic principles, would provide a useful framework to protect both resources and farmers' income.

Since its adoption, IPM (Integrated Pest Management) and its component Integrated Weed Management (IWM), has become the basis for all FAO plant protection activities because it contributes directly towards the achievement of sustainable agriculture in developing countries.

It is worth noting that IWM presents some important differences in relation to other IPM sectors:

1. The weed flora usually includes several species that contemporarily infest the same plot/field so, in practice, it is normally necessary to estimate the overall loss caused by the whole set of species instead of the effect of a single species. However, situations do exist in which the infestation is monospecific, or can be considered as such in relation to the competitive results (e.g. wild oat infestations in wheat, barnyardgrass in rice, gramineae in dicotyledon crops already treated in pre-or post-emergence with herbicides for controlling broad-leaved weeds).
2. Weeds have a strong periodicity, with plants co-existing at different stages of development; each species and each development stage within a species has a different impact and different sensitivity to the control measures, especially chemical weed control.
3. Herbicides do not generally control a single species but more than one, each with a different level of efficacy; although herbicides with a highly specific action spectrum exist. It is therefore not possible merely to extend to weed control the approach used for some time in the entomology sector, where each insect is controlled with specific products.

To implement an IWM strategy successfully, weed management should match the specific problems in a field and therefore some basic knowledge on weed and crop ecology and biology is needed to correctly predict the impact of a weed infestation on crop yield.

Within this context, weed-crop growth characteristics and the dynamics of weed emergence are important aspects. Many farmers in the developing world are unaware of several aspects of weed interference and the best time for weed removal, although there are exceptions. Ellis-Jones *et al.* (1993) found widespread recognition in

Zimbabwe of the importance of early weeding for both weed suppression and improvement of rainfall infiltration.

Weed germination patterns generally result in cohorts of seedlings emerging over an extended period of time and are heavily influenced by weather conditions, soil type and cropping system. The initial emergence time differs from year to year and varies according to the species ecological requirements (mainly temperature and soil moisture content-Forcella *et al.* 1997). It is also well established, and has been experimentally quantified for several crops and types of weed infestation, that the relative time of crop-weed emergence and the time of weed removal strongly ifluence crop production.

On small-scale farms in developing countries more than 50 percent of labour time is devoted to weeding, and is mainly done by the women and children in the farmer's family. In traditional farming systems knowledge of the so-called 'critical period' of competition would enable farmers to make the most efficient use of limited labour resources. In conditions of medium-high weed pressure, the critical period is approximately centred on the first one-third of the crop growing cycle.

For example, several major arable crops of temperate climates (e.g. maize, soybean, sunflower) take 100-140 days after emergence (DAE) to mature and the critical period is often between 25-40 DAE. Of course, the critical period varies with the relative competitivity of the crop and the weeds: the lower is the crop competitivity (and/or the higher the weed flora competitivity) the longer the period the crop must be kept weed-free to prevent significant yield losses.

Competition and Crop Yield

The presence of weeds in a crop leads to an increased number of plants within a certain area. Given that the crop density is already set at a level that optimises yield for that cultivar in that environment, the presence of weeds will lead to a reduction in the average yield of the crop.

In an infested field it is possible to identify different components of the overall competitive effect:

- intraspecific competition between plants of the cultivated species;
- interspecific competition between plants of the cultivated species and weed species;
- interspecific competition between plants of the different weed species;

- intraspecific competition between plants of the same weed species.

Predicting Yield Loss Caused By a Single Weed Species

Assuming that crop density does not vary significantly, i.e. that the effect of intraspecific competition between plants of the cultivated species can be considered as constant, a simplified situation can be analysed: the effect of a monospecific infestation on crop yield. The yield loss caused by the weeds can be predicted with an acceptable level of precision through the use of mathematical models. These can be empirical or mechanistic. Empirical models are based on mathematical empirical relationships between some important independent variables (e.g. weed density or cover, growth rates, time of weed emergence in relation to that of the crop) and a dependent variable (usually crop yield) the variations of which should be interpreted and predicted.

Mechanistic models are instead based on the growth processes (i.e. light interception, photosynthesis, dry matter partitioning between different plant parts of the competing plants (crop and weeds)); they are obviously much more complex than the former and require a good knowledge of the mechanisms and interactions involved in the crop-weed system and a large amount of data or information as inputs. These models have a limited direct practical use, but are good study and research tools from which much simpler empirical models can be derived.

In this chapter, only empirical models will be considered.

Many empirical models have been developed to describe, and possibly predict, the effect of weeds in crops, many of which are based on the relationship between crop yield loss and weed density. In an attempt to solve the problems linked to predicting yield loss in relation to weed time of emergence, alternative models have been proposed based on the relative leaf area of the weeds and crop.

Prediction of Yield Loss on the Basis of Weed Density

The model most widely used to describe the yield loss depending on the weed density is based on the rectangular hyperbola model:

$$Y_L = \frac{iD}{1 + \frac{iD}{a}}$$

where Y_L is the relative yield loss, D is the weed density, i is a parameter that represents the initial slope of the curve and a represents the maximum yield loss found with a very high weed density.

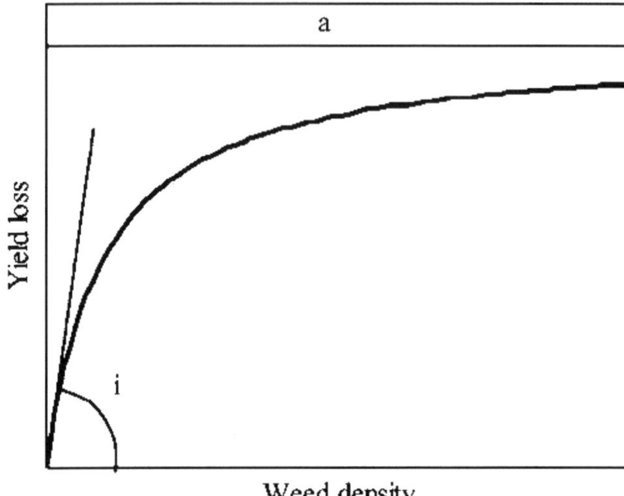

Figure: *Rectangular hyperbola (from Cousens, 1985a) that links the relative yield loss to the density of a weed species.*

Both parameters also vary with the crop-weed association as well as with crop density, time of emergence of the weed and crop and soil fertility. The values of i and a can therefore be used to compare various crop-weed associations in additive competition experiments (i.e. when the weed is added to a crop sown at a fixed density).

Equation (1) provides the relative yield loss, that cannot be measured directly, but only calculated starting from the yields observed with and without weeds. However, even the measurements taken in weed-free control plots can be affected by experimental error, so equation (1) can be adapted to include Y_{WF} among the equation parameters:

$$Y = Y_{WF}\left(1 - \frac{iD}{1 + \frac{iD}{a}}\right)$$

the third parameter Y_{WF} represents the yield of the weed-free control, which can thus be estimated using all the observed data and not just those from the control plots.

Equations (1) and (2) assume that the weeds emerge contemporarily with the crop, but in practice this does not happen. Cousens *et al.* (1987) therefore introduced a modification to the above-described model to take into account contemporarily the weed density and relative time of emergence of the crop and weeds:

$$Y_L = \frac{iD}{e^{c \cdot te} + \frac{iD}{a}}$$

where Y_L, D, *i* and *a* have the same meaning as in (1), te is the relative time of emergence of the crop and weeds and *c* is the regression parameter that expresses the variation of the competitiveness of the weeds depending on the delay with which they emerge.

Model (3) is an improvement on (1) as it allows the experimental data to be better described when the emergence is not contemporary with that of the crop, but with very gradual emergences the determination of *te* becomes difficult and time-consuming and the estimate of the parameters arguable.

Moreover, the estimate of parameters *i* and *a*, being dependent also on that of parameter *te*, generally show a higher variability over the years and with locations. This drawback can, at least partly, be obviated using the temperature sum instead of days as a 'biological scale'.

Weed density is the variable most commonly used to explain the variations in crop yield loss.

This has clear advantages, such as: a) simplicity of its control in experiments; b) determination in the field is relatively easy and fast. It also shows some disadvantages: a) in the field it is difficult to control the time of emergence of the weeds in relation to that of the crop, particularly when emergence occurs in flushes; b) the models based on it have weak eco-physiological bases.

For these reasons some authors have tried to identify another variable that would at the same time be easy to measure, take into account weed competitiveness determined by different times of emergence of crop and weeds, and be eco-physiologically sound.

Prediction of Yield Loss Based on Leaf Area or Relative Weed Cover

Kropff and Spitters (1991) proposed a relationship based on the relative leaf area (Lw) defined as the ratio between the leaf area index (LAI) of the weed and the total LAI of the crop plus weeds:

$$L_W = \frac{LAI_W}{LAI_c + LAI_W}$$

where LAI_w and LAI_c are the leaf area index of the weed and crop, respectively. L_w can vary from 0 (absence of weeds) to 1 (leaf cover of the weed alone). The authors, starting with the processes that regulate crop growth and the results of many simulations done with an eco-physiological competition model, demonstrate that the relative leaf area of the weeds compared to that of the crop at the moment of "row closure" could be a crucial measurement of the competitive process and very well correlated with the final yield loss.

The relationship between relative yield loss Y_L and L_w is expressed by:

$$Y_L = \frac{qLw}{1+(q-1)Lw}$$

where q is an index of competitivity typical of a given weed in a crop, called relative damage coefficient.

Kropff *et al.* (1995) modified equation (5) by inserting a parameter m that represents the maximum damage caused by the weeds (asymptote of the hyperbole):

$$Y_L = \frac{qLw}{1+\left(\frac{q}{m}-1\right)Lw}$$

Model (6) is comparable to (1) with relative leaf area instead of density.

Considering L_w instead of density takes into account, at least indirectly, that the damage caused by the weeds depends on the relative development of crop and weeds and therefore also their relative time of emergence. A certain value of L_w can be given by a few early-emerging plants or by many late-emerging ones.

The major problem of this approach is the difficulty in measuring L_w quickly and reliably. In fact, the precise measurement of leaf area is possible with destructive sampling of the vegetation (weeds and crop), but this method is neither rapid nor economical. An alternative is a visual survey, with the operator estimating the leaf area ratios between crop and weeds. With adequate training it is probably possible to obtain good results, but a level of subjectivity in the evaluations of L_w cannot be eliminated.

In place of L_w, the competitivity of an infestation can be evaluated on the basis of the partitioning of the upper layer of plant cover between crop and weeds (Relative Cover, RC). Observing the canopy vertically from above, the upper part of the plant cover will be formed by leaves of the crop and weeds: the ratio between the ground area covered by leaves of weeds and the total cover (i.e. weeds plus crop) represents the RC.

With W and C being the areas covered by weeds and crop, respectively, RC is given by:

$$RC = \frac{W}{W + C}$$

The higher this value is, the greater the share of solar radiation intercepted by the weeds will be, and therefore the competition caused by them will be more intense. This method assumes that: 1) interference for light is a measure of interference by all mechanisms: the leaf canopy may serve as an 'integrator' of the combined effects of competition for light, water and nutrients, and possibly also allelopathic effects, since these all reduce height, shoot weight and therefore leaf area and radiation interception; 2) the competitive effect of weeds that are shorter than the crop at canopy closure is negligible; in other words, only the plants that are able to overgrow or, at least, reach a height similar to the crop can successfully compete. The use of this variable was proposed with crops of medium-low height (e.g. peas, soybean): under these conditions, the use of the RC was shown to be valid. The applicability of this method still has to be evaluated with more widely spaced crops and on taller ones, such as maize and sunflower.

The ratio between RC and crop yield loss is similar to:

$$Y_L \frac{qRC}{1 + \left(\frac{q}{m} - 1\right) RC}$$

Both L_w and RC proved to be good yield loss predictors at canopy closure stage, which is far too late for any control treatment. The early estimation of these yield loss predictors involves estimating their evolution from the time of assessment to canopy closure. This represents a further source of variability that has to be determined by means of appropriate experiments.

The main advantage of RC is the easier measurement. While L_w is based on LAI ratios and therefore requires all the leaf areas to be determined, RC requires the measurement of the ratios of cover between crop and weeds by observing the canopy vertically from above.

This can either be done by means of a subjective visual estimate or, with greater precision, by measurements starting from photographs taken from 2-3 m above the crop. The possibility of using optical equipment that automatically take this measurement can also be foreseen.

It should be stressed that the basic data used for the determination of L_w and RC do not give any indication on weed flora composition, information that is fundamental for a correct choice of type of control.

Damage Caused by Mixed Infestations

Information on single crop-weed associations, although interesting, is of limited practical interest because infestations normally include various species and the control measures, in particular herbicides, have a well-defined spectrum of efficacy. The choices between control options (if and how to treat) must therefore be made, taking into account the damage that would be caused in the absence of control as well as the damage caused by the weeds that survive a given treatment. Both these evaluations require the competitive effect of a mixed infestation to be estimated.

In the case of the ratios between density and yield loss, the commonly used methods are based on the transformation of the observed densities into values that can be considered additive. It is obvious that two different weeds, even at the same density, will usually cause different yield losses: it is not therefore possible to directly sum up the observed density values to estimate the competitive effect of the infestation as a whole.

There are three main approaches: the first, proposed by Wilkerson *et al.* (1991) involves calculating the total competitive load (TCL). This method is used in the HERB program for the choice of post-emergence control options in maize and soybean. The second approach is based on the Density equivalent concept (Deq) and is the basis of a decision support system (GESTINF) adapted to Italian conditions and currently at the experimental stage at farm level. The third refers to relative cover and is still at the experimental stage.

Total Competitive Load Method

The competitive ability of the various weed species is assigned through competition experiments by using an indexing method, i.e.

assigning an arbitrary value K to the most competitive species and ranking the others according their relative competitivity with the reference species. Indexing is based on linear relationships between crop yield or biomass and weed density, which, for low density values, represent a good approximation of the hyperbolic relationship that exists between these two variables.

For low values of weed density (D), the crop yield can be given by:

$$Y = a + b\ D$$

The ratio b/a represents an index of the competitive ability of the considered species. A competitive index (CI) can therefore be defined for the i_{th} species, obtained from the following:

$$CI_i = \frac{b_i / a_i}{b_r / a_r} K$$

where b_r and a_r are the regression parameters of the reference species.

K is a scale factor that can assume any value; the authors have chosen a value of 10. In this case, the weeds are ranked following a decimal scale with CI=10 for the reference species. CIs are the basic data used for the evaluation of the yield loss in a real field situation. Then, multiplying the density observed of the i_{th} species by its CI, the competitive load (CL) for this species is obtained:

$$CL_i = CI_i\ D_i$$

The sum of these values for the various species present represents the TCL of that particular infestation:

$$TCL = \sum CL_i$$

Crop yield loss can be estimated on the basis of the TCL of the infestation. According to Wilkerson *et al.* (1991) the yield loss for low infestation levels can be calculated with a linear relationship. With the increase of the CL, the weeds start to interfere with one another as well as with the crop and the competitive effect caused by each single plant decreases. Under these conditions the crop yield loss follows a hyperbolic trend.

In the case of soybean, Wilkerson *et al.* (1991) fixed the passing from the linear relationship to the hyperbolic one at a TCL of 50. It is worth mentioning that this model was calibrated for the central-southern United States, where soybean was grown at wide interrow spacing.

The complete expression of the yield loss as a function of TCL will therefore be:

$$Y_L = \begin{cases} 0.5TCL & \text{for } TCL \le 50 \\ 25 + \dfrac{55(TCL-50)}{TCL+60} & \text{for } TCL > 50 \end{cases}$$

Density Equivalent Method

The Deq of a given weed species is defined as the density of a reference species that determines a yield loss equal to that caused by the studied species at the measured density.

The crop yield loss in competition with the reference species is:

$$Y_L = \frac{i_{ref} D}{1 + \dfrac{i_{ref} D}{a_{ref}}}$$

While for the i_{th} species present you have:

$$Y_L = \frac{i_i D}{1 + \dfrac{i_i D}{a_i}}$$

From the above definition, the Deq of the i_{th} species (Deq_i) is the value of the density of the reference species that makes the two preceding equations equal:

$$\frac{i_{ref} * Deg}{1 + \dfrac{i_{ref} * Deg}{a_{ref}}} = \frac{i_i * D_i}{1 + \dfrac{i_i * D_i}{a_i}}$$

A series of algebraic steps leads to:

$$Deg = \frac{i_i * D_i}{i_{ref} + i_{ref} i_i D_i \left(\dfrac{1}{a_i} - \dfrac{1}{a_{ref}} \right)}$$

The equation can be simplified choosing a hypothetical species as reference with the *i* and *a* parameters both equal to 1. This assumption gives:

$$Deg = \frac{i_i D_i}{1 + i_i D_i \left(\dfrac{1}{a_i} - 1 \right)}$$

Adding up the Deq of the various species present gives a total Density equivalent (Deq_t):

$$Deg = \sum Deg$$

The crop yield loss will then be obtained from:

$$Y_L = \frac{Deg_t}{1 + Deg_t}$$

Knowing the indexes *i* and *a* of each weed species it is possible to calculate the damage that can be caused by any combination of these weeds.

Relative leaf area and relative weed cover methods.

If the competitivity of the infestation is evaluated using the Lw instead of density, the equation that expresses the effect of a mixed population becomes simpler. The formula that is applied for a single species can in fact be developed in an additive way, giving the following:

$$Y_L = \frac{\sum q_i Lw_i}{1 + \sum \left(\frac{q_i}{m_i} - 1 \right) Lw_i}$$

where q_i, m_i and Lw_i indicate the relative damage coefficient, the asymptotic yield loss and observed value of Lw for the ith species, respectively. Where a maximum asymptotic value of yield loss is not considered, the value of m_i is equal to 1 for all the species.

When using RC instead of Lw as a descriptive variable for weed-crop competition, the extension to a multi-species situation is straightforward. RC *per se* measures the ratio between the horizontal projections of crop and weed leaves, so integrating the effects of both leaf area and leaf posture of different weed species. This implies that RC is intrinsically a multi-species descriptor. Measurements taken in the field can therefore be directly used for estimating yield loss caused by a mixed weed infestation.

Relationships Between Crop Yield Loss and Weed Time of Emergence and Removal

Given the importance of early growth, the trend of the relationships between yield loss and time of emergence and removal of weeds can easily be understood. The competitive effect of a given density of weeds emerging with the crop depends strongly on the length of the period they remain in the field (i.e. the time of weed removal). The relationship between the duration of competition and crop yield reduction is approximately sigmoidal: weeds competing for a short period have little effect on crop yield; allowing the weeds to compete for a longer time, the yield reduction increases, until a plateau is reached corresponding to the yield loss caused by weeds competing over the entire growing cycle. Crops like maize and soybean show a relatively long initial period when the damage caused by weeds is relatively low, while most horticultural crops are more sensitive, as shown in Figure below.

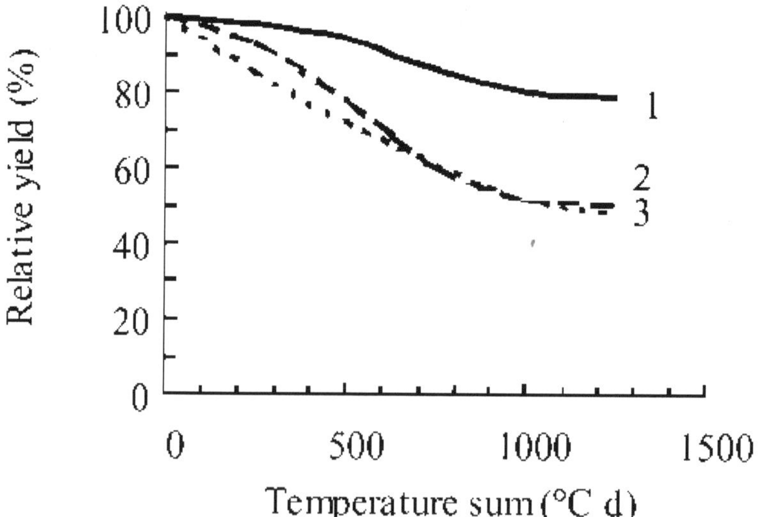

Figure: *Crop relative yield, as a percentage of the weed-free test, as a function of duration of competition (i.e. time of weed removal) for a weed density of 10 plants m^{-2}. 1 = maize in competition with Abutilon theophrasti (Sattin et al. 1992); 2 = soybean in competition with Amaranthus cruentus; 3 = onions in competition with Helianthus annuus.*

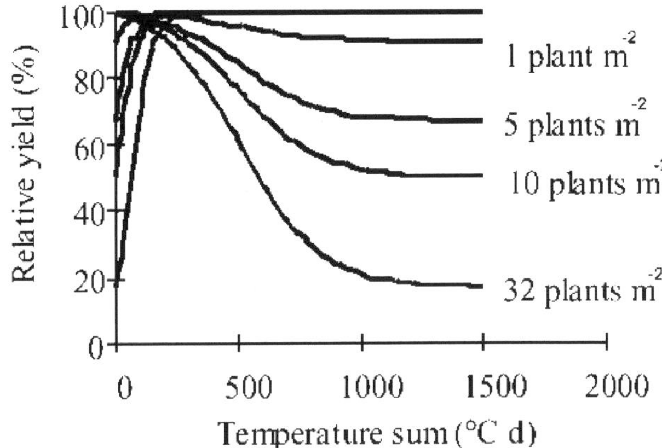

Figure: *Soybean relative yield, as a percentage of the weed-free test, in relation to time of emergence, time of removal and density of* Amaranthus cruentus.

The relationship between the time of emergence of weeds and crop yield loss mirrors the curves in Figure above. Of course these relationships are influenced by weed density: for a specific weed time of emergence or removal, the higher the density the lower is the relative yield.

Extending the above concepts to a mixed-weed infestation that can be found in any cropped field, the yield can be expressed as a function of maximum yield of the crop kept weed-free, weed-competitive load and time of emergence and removal of the weeds.

The first two factors are clearly site-specific, while contrasting data have been reported for the variability of the weed-free period (WFP) and duration of tolerated competition (DTC) curves for different years and/or locations. The experiments designed to obtain this sort of information are tedious and very expensive.

An important question is therefore how variable are these data in space and time. Using three data sets showing the effects of mixed weed infestations in maize, soybean and durum wheat and by means of the Deq approach, Sattin *et al.* (1996) analysed the variability of DTC and WFP curves over reasonably homogeneous areas in temperate environments. Despite the differences in weed flora among the experiments within each data set, the pattern of these relationships appeared to depend more on crop characteristics than on the composition of the weed infestation.

If these results are confirmed for other areas and crops, for a given area the prediction of yield loss caused by a mixed-weed infestation with any time of emergence and/or removal will require the knowledge of the weed-free yield, the relationship weed competitive load-yield loss and only one set of parameters (i.e. only one large experiment) linked to the effects of the time of weed emergence and removal. Information on weed competitive load-yield loss relationships already exists for several sites and crops and, if unavailable, it is much less tedious to obtain compared to the determination of the relationships between DTC, WFP, weed density and crop yield loss.

Critical Period and Optimum Time for Post-emergence Application

The critical period has been defined as the period during which weeds must be controlled to prevent yield losses. Since the concept of critical period was introduced, it has been used to determine the period when control operations should be carried out to minimise yield losses for many crops.

Historically, critical periods have been calculated by mean separations (hereafter referred to as the classical approach) in experiments that evaluated the impact of time of weed emergence and time of removal on crop yields. Using the classical approach, it is possible to identify a period within which no statistically detectable yield losses occur. It has also been concluded that for most field crops it is unnecessary to control weeds in the first few weeks after crop and weed emergence.

Several problems inherent to the classical approach have been pointed out and the use of regression analysis (hereafter referred to as the functional approach) was suggested as a better alternative. In changing from the classical to functional approach, the existence of a period when weeds do not cause any yield reduction became doubtful because of the continuous relationship between yield loss and time of weed emergence and removal.

To avoid this problem, fixed yield loss thresholds were used to define critical periods. Within this framework it was concluded that early weed control is unnecessary. This conclusion was not the result of an accurate evaluation of when to start to control weeds, but of the way critical periods were calculated.

The functional approach fails to recognise that since there is a continuous relationship between crop yield and time of weed removal, controlling weeds pre-plant or pre-emergence or post-emergence cannot be compared without considering yield losses that occur between planting and post-emergence control.

The establishment of a fixed yield loss threshold indirectly considers the economic aspect in the calculation of critical periods. Within this framework, Dunan *et al.* (1995) developed an economic approach to calculate critical period.

They defined the economic critical period as the time interval when the marginal income of weed control is higher than the cost of control, and its limits are called early and late economic period thresholds.

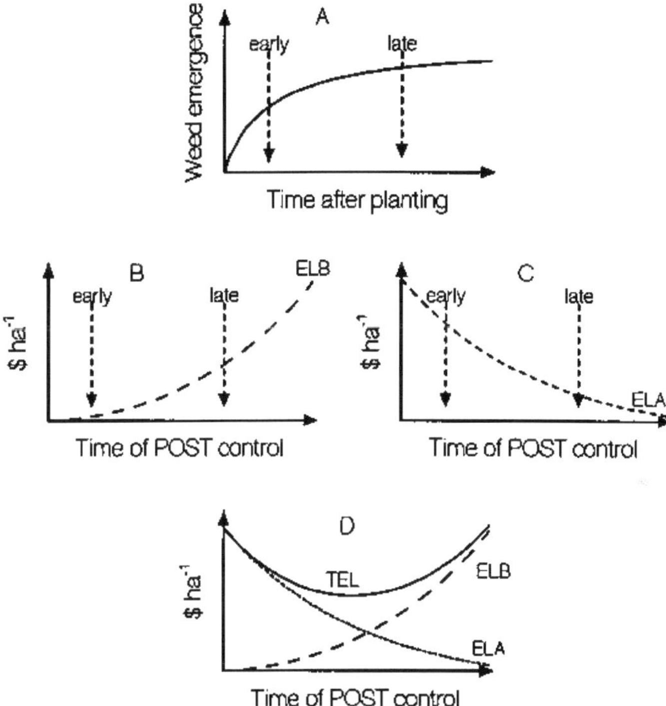

Figure: *Theoretical representation of the concept of optimum time of application for post-emergence treatments.*

(A) Hypothetical pattern of weed emergence as a function of time after crop planting.

(B) Economic loss as a result of weeds emerging before the post-emergence treatment (ELB).

(C) Economic loss as a result of weeds emerging after the post-emergence treatment (ELA); and

(D) Total economic loss (TEL) as a function of time of control. Early and late represent two different times of post-emergence weed control and it is assumed that the efficacy of post-emergence treatment is 100 percent.

The period when weed control can be carried out starts before sowing with a pre-plant weed control tactic and continues until the phenological stage of the crop, which precludes any further weed control. Within this time lapse the three approaches (classical, functional and economic) all define a period when weeds should be controlled.

However, identification of a period alone does not provide information on how many or when control practices should be performed. Considering the time dependence of the weed competitive effect, treatments carried out within this period cannot be regarded as having the same net margin (i.e. difference between the value of the crop with control minus the cost of treatment and the value of the crop without control) and therefore an optimum time for one or more weed control tactics exists which provides the highest net margin.

Weed emergence can begin from the moment of final seedbed preparation and then continue while environmental conditions are favourable for the germination process. As previously shown, the time of weed emergence greatly affects weed competitiveness, with the first emerging weeds being far more competitive than the late emerging ones.

Most post-emergence treatments have low or nil residual activity. Because of this they can control the weed population present at the moment of application, but have little or no effect on subsequent germinations. The economic result of post-emergence treatments will vary according to the time of application and is related to the efficacy of the treatment, crop characteristics (WFP and DTC curves) and weed germination pattern.

As a first approximation, crop yield loss is given by the sum of the yield loss caused by weeds emerging before the treatment (competing with the crop until the treatment) and that caused by weeds emerging after the treatment (competing from emergence until harvest).

These two yield losses follow opposite trends. With an early treatment, the loss caused by the weeds emerged before the spraying is low because the duration of competition is short. On the other hand, a consistent number of weeds can germinate after the treatment and, remaining until harvest, can produce an important yield loss.

With a late treatment, damage caused by weeds emerging after the treatment is reduced, but there is a marked increase in the damage as a result of the weeds emerging before the treatment. The sum of these two losses gives a total economic loss curve characterised by a minimum which identifies the optimum time of treatment.

Figure above shows an example of these calculations for maize in the Po valley-Italy and for soybean in southern Ontario-Canada. The curves were calculated using the Time Density Equivalent approach and the DTC and WFP relationships calculated by Sattin *et al.* (1996) were used. With the TDE approach, the weed population is divided in daily cohorts depending on their time of emergence. For each cohort the TDE is equal to the density of a cohort of weeds emerging with the crop and competing until harvest, giving the same yield loss as the considered cohort. TDEs can be considered additive and their sum is a Total Time Density Equivalent (TTDE), which gives a good estimate of the competitivity of the weed population depending on the emergence pattern and, considering herbicide efficacy, on the time of treatment. The relationship between time of weed control and net margin depends on the pattern of weed emergence and on the competitiveness of the crop.

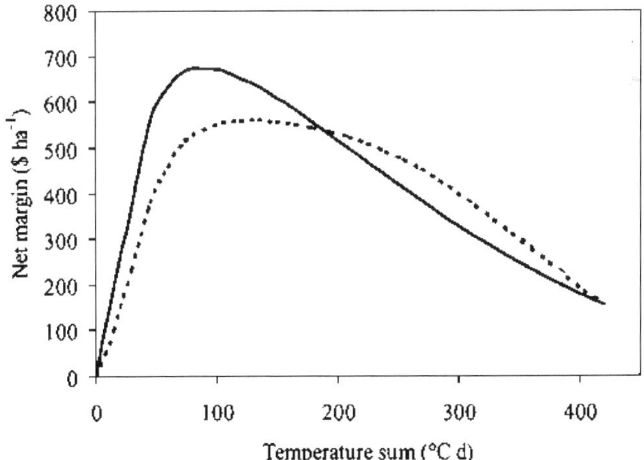

Figure: *Relationships between time of application of a post-emergence treatment and net margin for maize and soybean. Maize: weed-free yield = 10 t ha^{-1}; cost of post treatment = 45$ ha^{-1}; grain price = 155$ t^{-1}. Soybean: weed-free yield = 4 t ha^{-1}; cost of post treatment = 45$ ha^{-1}; grain price = 250$ t^{-1}. For both crops a weed infestation capable of causing a yield loss of 50 percent if untreated was considered.*

The Problem of Weed-pressure Assessment

The approaches presented above would allow to decide if, how and when to treat. This obviously requires some knowledge on the quantitative and qualitative composition of the weed flora. Depending on the approach used, different measurements should be taken (i.e. weed density, weed and crop LAI, relative cover). This phase of scouting is crucial: it is clear that the higher the level of precision the better it is, being the basis for subsequent calculations and for the selection of the weed control option to be adopted; on the other hand, scouting requires time and represents a cost, decreasing the net margin of weed control. Furthermore, weeds usually have a patchy distribution in fields and this requires the sampling techniques to account for this.

In the case of weed counts, Berti *et al.* (1992) proposed a relatively easy sampling method. Scouting is done by throwing a metal frame (25 × 30 cm) at random: the plants present within the rectangle are counted and classified by species. On the basis of previous studies, it has been verified that 20-30 throws are enough to guarantee a good estimate of weed density, assuming that the infestation is relatively homogeneous; if not sub-samples should be taken. The time required for this type of scouting is around 20 minutes per hectare when it is done by well-trained personnel, but can be more than doubled with less skilled people. The time requirement and the need for specialised personnel can become a major problem in the application of Decision Support Systems (DSS) in the field, especially in developing countries.

Other approaches have been proposed based on visual estimates of weed pressure. In the United States, Harvey and Wagner (1994) used a visual estimate of weed pressure for estimating crop yield loss. Weed pressure is defined as the visual estimate of the percentage which weeds contribute to the total volume of both crop and weeds in a plot, with a value of 0 indicating the complete absence of weeds and 100 indicating the complete absence of crop plants. A common problem with this type of approach is the high variability of ratings assigned by different personnel for the same level of infestation, leading to an uncertainty of yield loss estimates, particularly with low levels of infestation, which represents, on the other hand, the situation where the use of a DSS is more useful.

A further constraint is that the visual estimation of weed pressure is easily feasible when both crop and weeds have already reached an advanced stage of development, while it is very difficult to give a correct rating when plants are at the seedling stage. This limits the usefulness of this approach for the selection of present-year weed-control measures, while it might be used to predict future crop yield losses in the same field in order to assist in making sound weed-management decisions for the following year.

Therefore, at the present time the use of DSS for selecting weed-control measures faces a paradox: They are especially useful to less skilled farmers, because they can greatly improve their management techniques, while more experienced people can normally select the 'right' control measure based on their own experience. On the other hand, to be done rapidly and reliably, thus not determining an excessive increase of control costs, a proper assessment of weed competitive pressure requires a relatively high level of knowledge.

The use of models relating weed density or other independent variables expressing weed competitivity can greatly improve the weed control selection procedure, with a great impact on both yield and economics. However, applications of DSS derived from these competition models are, so far, relatively limited, mainly because the use of DSS requires a relatively high level of knowledge to be really effective.

The wide literature on this subject shows that the beginning of the growing cycle is crucial in determining the intensity and outcome of the subsequent weed-crop competition. It also appears that in hotter climates where the crop growing cycle is often shorter, as is found in many developing countries, the importance of early weed-crop competition is accentuated. The relative time of weed and crop emergence appears to be crucial and can overwhelm many other factors: a few days' difference in emergence can create an unrecoverable gap between plants, and therefore any agronomic practice which delays weed emergence and favours a satisfactory crop establishment plays an important role. There are indications that the variability (in space and time) of yield loss caused by mixed weed infestations can be relatively low. Therefore, if these results are confirmed for different areas, for various crops and for a reasonably homogeneous area, only a few low-tech experiments, although labour-intensive, are required to predict yield loss in relation to WFP, DTC and weed density.

This type of information could give useful indications and provide a framework particularly in situations where the technological level is low (i.e. where hand-weeding is prevalent), and then where simple rules directing weed control are more likely to be adopted instead of more complicated computer-based technologies. It would therefore be very useful to have, at least for the most important agricultural areas and crops, some basic data sets related to weed-crop competitivity, timing of weed competition and weather conditions.

7

Weed Control for Farm Forestry Plantations

This Agriculture Note provides an overview of the weed control required for successful establishment of farm forestry plantations. It includes information complementary to the Agriculture Notes AGO770: Site preparation for farm forestry, and AGO788: The use of fertiliser in farm forestry.

Weed control in farm plantations and woodlots is crucial during the first 2 growing seasons. Without adequate weed control, competition for water, nutrients and light can lead to high early mortality of trees. Slow growth of surviving trees is inevitable if weeds are not controlled.

Good weed control leads to fast early growth, and trees more rapidly dominate the site, providing their own measure of weed control by shading out weeds, particularly along the cultivated rows (ripped, mounded or both) or seeded rows. Good weed control also ensures that nutrient and water is available to the trees. Grasses in particular are very competitive along with some of the deeper rooted broadleaved weeds such as Flatweed-dandelion, Docks, Sorrel, and numerous others. Tall, deep rooted grasses such as Phalaris spp. can shade out trees as well as out-competing trees for water. Another example is Bracken fern-a severe competitor for phosphorus.

Weed control should be continued until the trees are dominating the site. For example, slower growing Pinus radiata (Radiata pine) requires 2 years of weed control. Generally the mean height of the faster growing Eucalyptus globulus (Blue gum), should be at least 2 m after 1 year. It is usually necessary to apply a second herbicide application in the autumn-early winter of the year after planting (second season weed control). Advantages often include better trafficability for machinery (sites can be too wet in winter – early spring) and minimal weed competition during the following spring-summer growth period.

Classification of Herbicides

Herbicides used in plantation establishment are usually systemic, that is, the chemical is translocated throughout the weed. These herbicides can then be described in terms of other properties as follows.

Foliar Knockdown Herbicides

These may be either broad spectrum ie. control a wide range of weeds, or specific to particular weed types eg. grasses, or classes of broadleaved weeds. These herbicides act only through foliar uptake by existing weeds.

Soll Residual Herbicides

Similarly, these may be either broad spectrum or weed class specific. They can act through:

- root uptake by existing weeds
- by pre-emergence or early post-emergence activity, such as contact by the emerging weed with chemical on the soil surface or close to the surface
- uptake by the developing roots
- some herbicides can be taken up by both foliage and roots.

Application of soil residual herbicides provides ongoing weed control for several months, particularly during the high growth seasons of spring and autumn. These herbicides require adequate soil moisture for activation.

Mode of Action

Herbicides are further classified as to their 'Mode of Action'. There are 14 Mode of Action groups with a letter code of A to N (found on the label). The Mode of Action designation indicates the point at which the chemical disrupts the plant growth chemistry. Examples of these disruptors include:

- photosynthesis inhibitors
- disruptors of cell growth
- inhibitors of protein synthesis.

Types of Herbicide Products

These may be liquids or solids. Liquids include flowables (which are very finely divided suspensions) and chemicals dissolved in solvents such as water, alcohol, or oils. Oil based products form emulsions when mixed with water.

Solids include dry flowables and wettable powders (which form suspensions when mixed with water) and water soluble granules (which are usually a high concentrate product).

A recent development is that of soluble packs of pre-mixed high concentration solids containing 2 or more residual herbicides. The herbicide mix is packaged so that a single pack is the correct quantity to broadcast spray a given area, usually one hectare. Another type is soil-dispersible granules, which contain only soil residual chemicals. These are best applied to bare soil and require soil moisture for activation (breakdown of the granule and release of chemical).

Site Assessment

Before any site preparation is carried out, the site should be assessed 9-12 months before planting for both weed cover and soil type. The weeds present may determine that immediate treatment is required.

For example, Bracken fern is very difficult to treat after trees are planted. This woody weed requires pre-conditioning for successful control, either by slashing or an initial herbicide application in the winter-spring to stimulate new growth. This should be followed by a further treatment in mid to late summer before site preparation. Other woody weeds may require treatment at a specific time of year eg. the appropriate time to treat Blackberry is December-January (at the early flowering stage).

Summer active grasses such as Phalaris, Couch, Paspalum or Kikuyu will also require treatment at an early stage, utilising the principle that weeds are best treated when small. Seed set control is also an option.

The type of soil is important when considering the rates at which some residual herbicides are applied either pre-or post-planting. In general, the lighter textured soils (sands) should be treated with lower rates of certain residual herbicides. However, a broadcast treatment is usually applied to sandy or light textured soils which are less water retentive, whereas a strip application may be appropriate for heavier textured loams and clay loams.

Non-chemical Weed Control

There are several mechanical and other non-chemical methods of weed control to consider prior to planting.

Pre-cropping

One approach is to grow a crop (excluding grass or grass cereal) on the site during the previous spring/summer. This eliminates annual weed seed set and improves tilth during site preparation, due to the initial cultivation. Hay cropping also removes seed and improves subsequent site preparation.

Scalping

Scalping is used in some circumstances for plantation establishment, for example, Pinus pinaster (Maritime pine) plantations in Western Australia. Scalping involves removing the weed and topsoil with a blade (tractor or bulldozer), and is usually only carried out on flat sites. Disadvantages include cost and possible tree nutrition problems resulting from the removal of topsoil. Further, on any land previously cropped, the soil seed bank will be present at depth, thus requiring deeper scalping.

Grazing

The weed mass can be substantially reduced by grazing prior to site preparation. Grazing can also be used to remove inter-row growth when the trees have attained a sufficient height eg. in the second spring and later when eucalypts have developed mature foliage and are in excess of 2 m tall, or later still in slower growing Radiata pine.

Other Non-chemical Methods

These are largely still experimental, and include heat kill of weeds with gas burner equipment set close to the ground (called flaming), or steam, which sterilises the surface soil, destroying near-surface seed. Weed mats are effective but add to cost.

Seed Set Control

The main agricultural weeds are grasses (annuals and bi-annuals) and broadleaves. These exist in high densities, up to thousands per square metre, and are severely competitive if not controlled. The survival of annual weeds is primarily dependent on massive seed sets every year, since the seed has a low dormancy period.

Seed set control involves 'freezing' the site before the weeds can set seed. 'Freezing' means the application of herbicides to prevent annual grass and broadleaved weed seed set at the growth stage just prior to elongation and formation of seed heads. It involves a broadcast kill of all weeds present, and differs from spray topping in agriculture where low rates of herbicides are used to prevent seed set without killing the grass.

Apart from reducing the quantities of all weeds in the following year, it selects weeds that have a greater reliance on seed dormancy. This means that there is a severe reduction or elimination of grass species, but an increased proportion of Clover, Flat Weed, Plantain and others. The result is a broadleaf dominant pasture, with a much lower overall weed cover. Ploughing before seed set is also effective. Grasses controlled early have a low lignin content, and residue breakdown and soil root release occurs far more quickly. The formation of large grass root clumps is prevented, so that cloddiness is reduced and tilth improved when ripping and mounding is carried out.

Timing of Herbicide Applications

It is important to maximise any accrued benefit from chemical weed control. Weeds are best controlled when small. Control at an early stage requires minimum label rates of chemicals, reducing costs and minimising the potential for harmful effects such as movement off-site.

A pre-planting application is usually used in eucalypt establishment, because there is no suitable immediate post-planting prescription. This also applies to most other non-Pinus sp. plantings and direct seeding. If ripping and mounding is carried out before the autumn break in the weather, and planting is in mid-late winter, there is normally a period of about 2-3 months between site preparation and planting which allow mounds to settle (and eliminate air pockets), during which weeds can germinate. However, ripping and/or mounding is not always necessary. Pre-planting applications of herbicide are carried out within a period of a month to about 10 days before planting.

Where there is existing weed cover on the planting rows prior to planting, tank mixes of a broad spectrum foliar knockdown herbicide with one or more soil residual herbicides are commonly used. Such tank mixes provide knockdown of the existing weed cover with the foliar knockdown herbicide, and maximise the soil residual period of the residual herbicides. For bare soils, the foliar knockdown component is not required.

Soil applied granular herbicides are not used for the initial establishment of eucalypts, but may be used in establishing Radiata pine.

For various reasons, weed control may not be sufficiently effective and a treatment in the late spring may need to be applied to control, for example, an emerging summer active grass. Suitable herbicides are available but are more expensive, and in some cases risky, because higher temperatures and good soil moisture may predispose actively growing trees to be more susceptible to phytotoxic (herbicide) damage.

For Radiata pine, there are suitable spray and granular herbicides which allow treatment to be delayed until up to 2 months or more after planting. Pre-planting treatment may be preferred, and is usually cheaper.

Table: *Plantation establishment sequence*

Time of year	***Action required***
October-November	Inspect weeds and determine soil type. List weeds, including annual/perennial grasses, Sorrel, Dock, Thistles, Ragwort, Blackberry, Bracken.
Before grasses flower, late October-November	Boomspray with appropriate tank mix to kill as much of the weed cover as possible.
December-February	Spray Blackberry during early flowering. Respray problem areas, areas of initial poor control, and re-emergent Bracken. Grasses should break down rapidly due to low lignin levels.
February-April	Ripping and mounding.
May-June	Following the autumn break, assess weeds on mounds, and spray with tank mix of foliar knockdown and residual herbicides.
July-August	Plant trees. Regularly inspect thereafter for weeds.

The above sequence is more suited to large plantations. As soil type, climatic conditions, locality and the type and size of plantations vary from one site to another, each site may require a different sequence. Spring planting or seeding is preferable on sites which are water logged in winter eg. basaltic soils in southwest Victoria, or duplex soils in Gippsland. The sequence then requires pre-planting application of herbicides in late winter or early spring. A broad spectrum foliar knockdown herbicide combined with soil residual herbicides kills existing or emerging weed and provides longer term control, preferably through to the following autumn. However, rates of application of herbicides are usually less than in the sequence.

This is because the period required for extended weed control (to the following autumn) is shorter, and because rates at which residual herbicides are applied are seasonally dependent. There is more risk of phytotoxic damage to actively growing seedlings if residual herbicides are applied late in spring or into early summer. Late winter to early spring applications prior to spring plantings, and use of lower rates for spring plantings and direct seeding, usually need to be considered in order to avoid phytotoxic effects.

Second Season Weed Control

Second season weed control is usually carried out in the autumn to early winter period of the year after planting, but may be delayed until spring if the initial first season weed control has given an effective weed free year.

Spray applications for eucalypt plantations are usually directed ie. half of each mound is sprayed using directed (sideways) nozzles on either side of a low boom. The herbicide prescriptions again usually include a foliar knockdown component which if sprayed over the top of the trees can cause unsightly short term scorch.

It may be necessary to also control inter-row weeds, either by slashing or by a herbicide application. Certain herbicides used for this purpose need to be applied by shrouded boom to protect the trees. Shrouds for boom sprays are usually heavy skirts on a frame over the boom, and can be of heavy plastic or 'straw broom' construction. Their purpose is to prevent spray drift of chemicals which will scorch the trees.

In the second season, Radiata pine can be oversprayed with appropriate products. For spring plantings or direct seeded species, early autumn overspraying after rain with minimum rates of residual herbicides is usually sufficient to provide weed control until the following spring.

For both eucalypts and Radiata pine, granular herbicide products can be accurately applied in the second season. Spot treatments of diameter 1-2 m are applied around a tree by the hand-held 'Weed-A-Metre'. Strip (band) treatments also use specialised granule applicators, which include a powered back-pack applicator and a tractor applicator (3 point linkage).

For strip applications, either spray or granular, the width should be at least 2 m, and on sandy sites, broadcast application may be preferred..

Dr Rod Bird and associates have carried out considerable work, regarding the establishment of direct seeded trees on farms, at the DPI's Pastoral and Veterinary Institute at Hamilton. Research has shown that overspraying of mixed species with many different herbicides in many cases increased survival. Although this is potentially harmful to the trees, the effects were not severe enough to cause significant damage.

Application Volumes for Spraying

To some extent, application volumes for spraying depend on the terrain. However, on pasture sites volume is usually in the range 100-150 L of spray mix per hectare.

Costs

Chemical cost of pre-planting treatment in the establishment of eucalypts, is less than $100 per hectare on a broadcast basis. Costs may be less that $50 for a strip application, depending on the proportion of the area treated (which in turn depends on the inter-row spacing). Costs are further reduced for applications for direct seeding or spring plantings, where lower rates are used. Chemical costs for establishing Radiata pine are higher if certain standard prescriptions are used. These 'standard' prescriptions are of long standing and may be spray or granular, but specific rates of one or both of two particular chemicals are usually applied. The costs of granules for both eucalypts and pine, are about $200 per hectare or greater, on a broadcast basis.

Weed Management in New Tree Plantations

Weeds are some of the worst enemies of young tree plantations. Unchecked weed growth will weaken young trees, slow their growth and often kill them. Until recently, most growers of forest plantations relied on herbicides to keep their plantations weed-free. However, restrictions on the use of herbicides and worries about their environmental effects are causing tree planters to adopt a new approach to weed control, "integrated vegetation management."

What is Integrated Vegetation Management?

Integrated Vegetation Management integrates several practices in combination to minimize weed competition. It relies on tilling, cover crops and mowing to control most of the weed growth; herbicides are used selectively around the tree seedlings themselves. Experiments are also underway to test the value of mulches and protective tubes.

How Does it Work?

Lets look at an example: in late summer, a grower plows down an old alfalfa/brome grass hay field and disks, drags and seeds the entire field to a perennial grass cover crop. The following spring, the farmer plants hardwood and conifer seedlings in rows and treats a two-foot band of soil in each row with a pre-emergent herbicide. The weed-free band is maintained and the grass between the rows is mowed once or twice per year until the crowns start to touch each other.

This integrated approach to weed control is what is meant by integrated vegetation management.

How Can I Control Weeds?

Tilling can often be used over an entire planting area on level ground, or in contour bands in erodible terrain. A single moldboard plowing will kill alfalfa and most perennial grasses. (Quack grass may require a year of summer fallowing.)

Disking between the tree rows can be an effective means of controlling weeds after planting. If the trees were originally planted in a square grid, disking can be done in both directions.

Cover Crops protect the soil and prevent invasion by weeds.

- Timothy is inexpensive to plant, stays reasonably short and can be mowed once a year, after it forms a seed head.
- Perennial ryegrass stays reasonably short and can be mowed often. However, it is expensive to plant, is subject to winter kill, and low, frequent mowing often allows invasion by dandelions.
- Orchard grass is inexpensive to plant, stays relatively short and is moderately shade tolerant.
- Tall fescue and red fescue are often used as cover crops but may be allelopathic to black walnut roots.

Mowing is a well-known weed control method and is effective between rows. However, because it does not kill the weeds, the grower must monitor weed growth regularly and mow often.

Mulches may be applied with varying success. Plastic mulches are difficult to keep in place and are not very long lasting. Shingles are long lasting but can be blown around by strong winds. Wood chips and sawdust are sometimes effective for a season or two if perennial weeds have been removed first. Protective tree tubes are in the demonstration stage, but are expensive.

Herbicide restrictions change frequently. For example, the forestry section has been removed from several simazine labels.

Oust is now registered for preplanting and post-planting application in over the top of several tree species. Keep abreast of new products and new pesticide restrictions by checking with your local DNR forester or a private forester.

Using Integrated Vegetation Management, a grower can achieve effective and environmentally-safe weed control. It requires careful planning, but your efforts will pay off with a well-established, healthy tree plantation.

Of special note:

- Perennial grasses are among the worst enemies of young walnut trees, through the sapling stage. Heavy brome and quack grasses overtop and smother newly planted trees, and heavy bluegrass sod often causes stagnation of sapling walnut plantations.
- Alfalfa is difficult to kill after trees are planted. Once the trees are planted, the only control method available may be hand application of herbicides. A separate leaflet is available from the Forest Pest Management Unit, Wisconsin DNR that details alfalfa control.

Management Options and Perspectives

Cover crops are plant species that are introduced into crop rotations to provide beneficial services to the agro-ecosystem. Some of the most important environmental services provided by cover crops include soil protection from erosion, capture and prevention of soil nutrient losses, fixation of nitrogen by legumes, increase in soil carbon and associated improvements in soil physical and chemical characteristics, decrease in soil temperature, increase in biological diversity including beneficial organisms, and suppression of weeds and pests. This chapter will focus on weed suppression by cover crops, but the need to manage cover crops to optimize the totality of impacts on the ecosystem will be emphasized at the conclusion.

Cover crops can be grouped into two categories: 1) annuals that are grown during an off-season or period of the year that is not favourable for crop production and that are killed before planting a cash crop; and 2) living mulches that grow at the same time as the cash crop for all or a portion of the growing season. Cover crops that are killed before planting a cash crop influence weed control primarily through the influence of their residue on weed germination and establishment.

Examples of this kind of cover crop are *Vicia villosa* Roth, a winter annual legume, and *Secale cereale* L. a winter annual cereal, which are adapted to grow during the cold season in temperate climates and are killed before planting a cash crop when temperatures become warmer. Examples of cover crops that are adapted to hot summer fallow periods in tropical and subtropical climates include annual legumes such as *Mucuna* spp. and *Crotalaria juncea* L. or warm-season annual grasses such as *Sorghum* spp.

Weed Control By Cover Crop Residue

Annual cover crops are usually killed before planting a cash crop. This can be performed either by incorporation of cover crop residue into the soil or by killing the cover crop chemically or mechanically and leaving the residue as a mulch on the surface of the soil.

Incorporated Residue

Tillage has been shown to stimulate weed germination and emergence of many weed seeds through brief exposure to light. When tillage is used to incorporate residue, many weed seeds will be stimulated to germinate by this operation. Therefore, when incorporating residue by tillage, weed management tactics must be available to control the potential increased load of weed seedlings.

Incorporated plant residues can become toxic to weeds by the release of allelopathic chemicals. There are numerous reports of allelopathy and of the isolation of allelopathic compounds from plants. However, this phenomenon can be inconsistent under natural conditions because the allelopathic potential of plants is affected by many factors such as the age of the plant, soil properties, and environmental conditions. Interactions of multiple stresses in the environment on the target plants also will affect the degree of allelopathic activity. Examples of successful weed control leading to an increase in crop yield following incorporated cover crop residue include incorporation of *Sorghum bicolor* L. stalks before *Triticum aestivum* L., incorporation of *Brassica napus* L. before *Solanum tuberosum* L., and incorporation of *Trifolium incarnatum* L. before *Zea mays* L..

As with the use of herbicides for the control of weeds, there must be sufficient selectivity between the activity of cover crop toxins on weeds and on crops. In order to be useful as a weed control practice, the crop must be relatively insensitive to allelochemicals in the environment. Small-seeded crop plants may be more sensitive to allelochemicals than large-seeded plants. Crop cultivar selection and appropriate residue management may be important approaches for maximizing allelopathic activity on weeds and minimizing deleterious effects on crops, including autotoxicity. The relative timing and placement of residue relative to crop seeds can be manipulated to reduce the level of toxicity that emerging crop seedlings are exposed to.

Surface Residue

When cover crops are killed and residue is left on the soil surface in a no-tillage cropping system, many factors will contribute to weed suppression. Absence of tillage itself lowers weed emergence because seeds that require a brief exposure to light during tillage operations are not induced to germinate. In addition, residue on the surface of soil can suppress weed emergence directly. The degree of weed control provided by cover crop residue on the surface of soil can vary according to cover crop species, residue biomass, and weed species. Weed suppression by cover crop residue increases according to a negative exponential relationship with increasing residue biomass. Residue levels that are naturally produced by cover crops can reduce weed emergence up to 90 percent. Annual species that are small-seeded and have a light requirement for germination are most sensitive to surface residue, whereas large-seeded annuals and perennial weeds are relatively insensitive. Weed suppression will decline during the course of the season as the residue decomposes.

Residues on the surface of soil can vary greatly in dimension, structure, distribution pattern, and spatial heterogeneity. Several physical properties of mulches have been explored that may contribute to weed suppression by the physical impedance of weed emergence. The 'mulch area index' is a pivotal property for defining many important mulch properties. It is defined as the projected area of mulch material per unit soil area and can be determined by multiplying the residue mass per unit area by the area-to-mass ratio as measured from a sub-sample of residue material. The 'solid volume fraction' is another important mulch property that is defined as the fraction of mulch volume composed of solid material. Together, these two indices can predict weed suppression by a wide variety of mulch types ranging from *Z. mays* stalks with a low area-to-mass ratio to *Quercus* leaves with a high area-to-mass ratio. This suggests that residue with a large number of layers and a small amount of empty internal space will be most suppressive.

Residue also influences the microclimate of the soil by intercepting incoming radiation. Interception and reflection of short-wave radiation by residue reduce the quantity of light available to the soil surface, the heat absorbed by soils during the day, and the amount of soil moisture evaporated from soils. These effects can interact with seed germination requirements to determine the pattern of weed seedling emergence observed in any given season.

Light extinction by cover crop mulch follows a similar negative exponential decline in relation to mulch area, as light extinction by a crop canopy declines as a function of leaf area. Since mulch mass is linearly related to mulch area, a similar exponential relationship holds between light extinction and mulch mass. Many weed species require light to activate a phytochrome-mediated germination process prior to emergence. Emerging weeds also require light for initiation of photosynthesis before seed reserves are depleted. Extinction of light by residue can be an important factor inhibiting weed emergence through residue.

Cover crop residue on the soil surface can reduce maximum soil temperature by 2-5° C and raise minimum soil temperature by 1° C in temperate climates although this will vary according to radiation intensity, soil moisture, and soil type. Greater differences will probably be observed in tropical or drier areas of the world. Most weed seed will germinate over a wide range of temperatures and, therefore, a reduction in maximum soil temperature by residue will usually have little influence on germination. Because of the decrease in maximum and increase in minimum soil temperature, the daily soil temperature amplitude also is reduced by residue. High temperature amplitudes often are required to break the dormancy of selected weed species and, therefore, a reduction in soil temperature amplitude by cover crop residue can prevent germination of weed species that have this requirement.

Residue on the soil surface increases soil moisture by increasing infiltration of rainfall and by decreasing evaporative moisture loss.

Higher soil moisture under cover crop residue could either benefit or retard weed germination, depending on species requirements. Under saturated soil conditions, residue could slow evaporation and reduce germination of species inhibited by excess soil moisture. Under droughty conditions, retention of soil moisture could enhance weed germination and seedling survival.

Residue in most fields will have a relatively heterogeneous spatial distribution. This can be caused by relatively uneven stands of cover crop plants within a field resulting in areas with locally heavy and thin residue after cover crop desiccation. Even when there are relatively uniform stands of cover crops, uneven residue at a microsite level can be detected. For example, greater than 50 percent of sites measured under seemingly uniform *Vicia villosa* mulch permitted greater than 10 percent light transmittance to the soil level. This can be explained by the exponential relationship between soil cover and mulch area index.

Assuming a random distribution of mulch material, it will require increasingly more mulch to achieve each successive unit increase in soil coverage by mulch. For example, it takes an increase in mulch area index from 1.4-1.9 (= 0.5) to raise soil cover from 75 to 85 percent but it takes an increase from 1.9-3.0 (= 1.1) to raise soil cover from 85-95 percent. Even a relatively high mulch area index of 4 will leave 2 percent of soil uncovered. Thus, cover crop residue will rarely provide complete ground cover and cannot be expected to provide either complete or full-season weed control. Cover crops can contribute to weed control but herbicides or other weed control tactics are required to optimize weed control and crop yield.

Every weed control tactic, including cover crops, exert a selective pressure on weed populations and will select for those species that are best adapted to that system. Perennial and selected large-seeded annual weeds that have minimal requirements to break seed dormancy and sufficient energy reserves to penetrate mulch layers will be most likely to establish and reproduce in a cover crop mulch. Also, species that have a similar phenology to the cover crop but that can survive the cover crop management system will become problematic. For example, we have observed that *Lolium multiflorum* Lam. can become established with a *V. villosa* cover crop or *Digitaria sanguinalis* (L.) Scop. can establish with a spring planted *Glycine max* (L.) Merr. cover crop and both species can regrow and reproduce after the cover crop is mowed in preparation for planting a cash crop. Thus, cover crops must be used in rotations that prevent the buildup of species adapted to that cover crop system.

Practical Applications

Cover crops that produce high amounts of biomass will enhance weed suppression by leaving high amounts of suppressive residue.

Vigorous species that are well adapted and planted at optimum planting dates will be most useful. For example, *Vigna unguiculata* (L.) Walp. is adapted to hot, dry conditions and produced 8.2-9.6 Mg/ ha of residue as a cover crop that effectively suppressed weeds in a desert climate. Mixtures of cover crops that have complementary resource requirements is another approach to increasing cover crop biomass. Often, a combination of grasses and legumes make effective cover crop mixtures for the same reasons they make effective intercropping partners. A polyculture of *V. villosa* plus *T. incarnatum* plus *S. cereale* produced higher biomass and suppressed weeds more than each species in monoculture.

Cover crop residue that decomposes slowly will extend the period of weed suppression.

Slow decomposition is associated with residue material that has a high carbon-to-nitrogen ratio. For example, residue of *S. cereale* which has a higher carbon-to-nitrogen ratio than the legume *V. villosa* had a more extended period of weed suppression than *V. villosa*. Also, equipment such as a mower that shreds residue would enhance decomposition compared to equipment such as a roller that keeps the residue intact.

Low amounts of residue can stimulate weed emergence.

Occasionally, more weeds will emerge in low levels of cover crop residue (1-2 mg/ha) than in uncovered control plots. Low levels of residue are not sufficient to inhibit weeds from emerging but can create an environment more favourable for germination and emergence. This residue can retard evaporation of soil moisture and provide more uniform moisture conditions for germination and emergence than exists at the surface of bare soil. Also, nitrogenous compounds released into the germination zone, particularly from legume cover crops, can stimulate germination of selected weed species.

Creating a mulch with multiple layers of densely packed material.

Since mulch area index and solid volume fraction are important determinants of weed suppression, management practices that create the maximum mulch area and solid volume or, conversely, that minimize empty mulch volume, will maximize weed suppression. Mulch composed of broad leafy material held in a matrix of grass stems as might be obtained from a legume-grass cover crop mixture may be more effective than mulch composed primarily of stems or leaves alone. Also, use of implements such as rollers or stalk choppers that pack or compress the mulch as part of the desiccation process may maximize the suppressive potential of cover crop mulches. The use of cover crops that will provide uniform stands and minimize gaps is recommended. This will maximize the area with optimum amount of residue and minimize the area with ineffective or stimulatory levels of residue.

Living Mulches

Living mulches are plants grown with a cash crop. They are usually not grown for harvest or direct profit but, instead, to provide ecological benefits including protecting soils from erosion, improving soil fertility, providing traffic lanes, suppressing weeds, and reducing pest populations. Low-growing legumes and grasses are typically used for this purpose. Forage and turf species often are used as living mulches because their growth habit is lower than most crops and they are relatively easy to establish and manage. Legumes are often included in cropping systems where improvements in soil fertility and quality are a primary goal, whereas grasses are often included where durability and traffic ability are important. Living mulch such as *V. unguiculata* or *Mucuna* spp. also may produce edible parts that can supplement income produced by the primary crop with which it is intercropped.

Established living mulches can protect crop plants by forming a barrier to weed and other pest organisms originating in soils. Living mulches also create a more diverse community that can reduce insect pest levels by attracting natural enemies of pests or by creating an environment that is more difficult for pests to find and multiply on crop plants. The major constraint to using living mulches is competition for water and nutrients leading to reduced crop yield. Creative management approaches are required to alleviate the detrimental effect of living mulches on crops while enhancing the benefits to weed and pest management.

Weed suppression by living mulches and the problem of selectivity.

Because weed and living mulch plants compete for the same resources, weeds can be suppressed by the introduction of living mulches into cropping systems. If a cover crop becomes established before the emergence of weeds, then the presence of green vegetation covering the soil creates a radiation environment that is unfavourable for weed germination, emergence, and growth. Several requirements for breaking dormancy and promoting germination of weed seeds in soils (light with a high red-to-far red ratio and high daily soil temperature amplitude) are reduced more by living mulches than by desiccated residue. Once established, living mulch also can use the light, water, and nutritional resources that would otherwise be available to weeds. Allelopathy is another mechanism by which living mulches may suppress weeds. However, this is difficult to separate experimentally from mechanisms relating to competition for growth resources. Weeds can escape suppression by living mulches through gaps in the mulch canopy, by morphological and physiological capabilities to access resources despite the presence of competitive living mulch, or by emergence and growth patterns that avoid the most competitive period of living mulch growth.

Cover crops that grow during periods when crops are not present in a rotation can aid in maintaining ground cover and occupying a niche that would otherwise be occupied by weeds. For example, cover crops planted in the fall provided a ground cover that protected the soil from erosion and suppressed weeds during a summer fallow in the Canadian prairies. In addition, cover crops planted in the fall can become living mulch for a crop that is relay-planted into the living cover crop in the following year. Enache and Ilnicki (1990) developed a system whereby *Trifolium subterraneum* L. was initially planted in the fall and produced a cover of dense, low vegetation that remained alive until natural senescence several weeks after corn was relay-planted in the spring. The subsequent mulch continued to suppress weeds throughout the remainder of the season until volunteer *T. subterraneum* emerged in the fall and established a naturally recurring cover crop. Weed biomass was reduced by 53-94 percent by this living mulch whereas weed biomass in desiccated *S. cereale* mulch ranged from an 11 percent decrease to a 76 percent increase compared to a no-mulch control. Likewise, weed biomass was reduced by 52-70 percent in live *V. villosa* treated in a manner similar to that described for *T. incarnatum* whereas weed biomass ranged from a 41 percent reduction to a 45 percent increase in desiccated *V. villosa* residue compared to treatment without a cover crop. Thus, a living cover crop is capable of greater weed suppression than desiccated cover crop residue.

Living mulches also can be intercropped with a primary cash crop by planting shortly before, at the same time as, or shortly after planting the primary cash crop. These secondary intercropped species are often referred to as smother crops. Smother crops should be species that establish more rapidly than weeds and whose peak period of growth coincides with that of early weed emergence but does not coincide with that of the crop. Ideally the smother crop should suppress weed establishment during the critical period for weed establishment, i.e. the period when emerging weeds will cause a loss in crop yield. The smother crop will then senesce following this critical period for weed competition, thereby minimizing subsequent competition between the smother crop and the primary crop during the remainder of the season. One approach is the use of low-growing, fast-establishing, fast-maturing annuals planted with longer-season grain crops. For example, using annual *Brassica* and *Medicago* spp., Buhler *et al.* (2001), observed various levels of weed control depending on seasonal, species, and timing variables. However, good control of weeds was usually associated with crop yield loss.

In tropical systems, Chikoye *et al.* (2001), planted several smother crops with various growth habits in a *Z. mays-Manihot esculenta* Crantz intercrop system and found that *Mucuna cochinchinensis* (Lour.) A. Chev., *Lablab purpureus* L. and *Pueraria phaseoloides* (Roxb.) Benth. were effective for reclaiming fields heavily infested with the difficult-to-control perennial weed, *Imperata cylindrica* (L.) Beauv. After three years, rhizome biomass of *I. cylindrica* was reduced by 94 percent by annually weeding five times, 89 percent by *M. cochinchinensis*, 77 percent by *L. purpureus*, 74 percent by *V. unguiculata*, and 55 percent by *P. phaseoloides*. Akobundu *et al.* (2000) observed that *Mucuna* spp. suppressed *I. cylindrica* until the subsequent cropping season when *Z. mays* yield was higher and hand weeding was reduced by 50 percent compared to plots without cover crop. *Mucuna deeringiana* (Bort) Merr. and *Canavalia ensiformis* (L.) DC. living mulches reduced weed biomass and improved Z. mays yields in a traditional slash-and-burn system in Mexico (Caamal-Maldonado, 2001).

Liebman and Dyck, (1993) reviewed literature where one or more primary crops were intercropped with a smother crop and found that weed biomass was lower with than without the smother crop in 47 cases, variable in three cases, and higher in four cases. Thus smother crops can be effective tools for managing weeds as well as improving soil fertility and providing additional food production if edible reproductive parts are produced by the cover crop.

The major hurdle to the adoption and use of living mulches is lack of selectivity. Typically, a living mulch that is competitive enough to suppress weeds will also suppress crop growth and yield. Much of the research with living mulches has focused on documenting and alleviating this problem. Several approaches have been used to reduce competition between the living mulch and cash crop species without eliminating the desirable attributes and benefits of the living mulch. These attempts to achieve selectivity have met with varying success but often lack consistency.

Practical Applications

Ideal living mulch for weed suppression should have the following characteristics:

- ability to provide a complete ground cover of dense vegetation;
- rapid establishment and growth that develops a canopy faster than weeds;
- selectivity between suppression of weeds and the associated crop.

Means for achieving selectivity between weeds and the associated crop include:

1. Using low-growing living mulch that competes primarily for light. In this case, as long as the living mulch becomes established before the weeds, it would maintain weed suppression by excluding light but would not impact taller growing crops and would not compete with the crop excessively for soil resources such as water and nutrients.
2. Planting the living mulch so that the time of peak growth of the living mulch does not coincide with the critical period during which competition would have the greatest impact on crop yield.
3. Reducing crop row spacing and/or increase crop population to enhance the competitiveness of the crop relative to the living mulch.
4. Providing supplemental water and nitrogen to compensate for resources used by living mulch plants.
5. Suppressing the living mulch so as to reduce its competitiveness with the crop.

Means for suppressing living mulch include:

a) A broadcast application of an herbicide at a rate that is suppressive but not lethal.

b) A banded application of a herbicide to kill the living mulch in the crop row so as to reduce competition within the row area but permit weed suppression by the living mulch between rows.

c) Strip tillage to provide suitable planting conditions without competition within the crop row but to permit weed suppression by the living mulch between rows.

d) Mowing to reduce the height and vigour of the living mulch.

Cover Crops as Part of an Integrated Weed Management System

Holistic management principles and a shift to a systems approach for crop protection are vital to combating agricultural weeds as well as other pests. Ecologically-based weed management focuses on preventive practices and natural processes of population regulation with herbicides or cultivation used as interventions only when needed. Emphasis is placed on maximizing the beneficial ecological processes within farming systems that can maintain weed populations at low, manageable levels. Although agricultural systems are simplified compared to natural ecosystems, there are abundant opportunities to redesign and manage agricultural systems to reduce weed populations.

The live and dead plant materials associated with the use of cover crops in agricultural systems are particularly well suited to developing ecologically-based weed management systems. Generally, a more diverse biological and physical environment at the surface of soils such as that associated with cover crops offers opportunities for regulating and minimizing weed populations. Liebman and Gallandt (1997) propose that successful integrated weed management systems can be developed by combining several strategies or 'little hammers' that would cumulatively reduce the relative fitness of weeds versus crops.

An integrated system, including cover crops in combination with other strategies, could improve weed control compared to reliance on each strategy alone. Not all weed management strategies are equally compatible with cover crops, however. For example, soil-active herbicides can be adsorbed by cover crop residue and are less effective with than without cover crops. Mechanical cultivation is often not as efficient in reduced tillage systems where living and/or dead cover crop vegetation can interfere with cultivating equipment and where untilled soil is less susceptible to fragmenting and desiccating weed seedlings as is a clean, well-tilled soil. Cover crops should be more compatible with control measures such as post-emergence herbicides or biocontrol agents that act on the foliage of weeds after emergence than practices that act through the soil medium. Most important, long-term strategies need to be developed to maintain weed populations at low levels through suppressive crop rotations, crop population/row spacing, and fertility management.

Ultimately, weed management is one of the many potential benefits of using cover crops. Cover crop management therefore must be designed to optimize all of the potential benefits that can be derived from cover crops and to minimize the negative impacts of cover crops.

For example, high levels of cover crop biomass may be desirable for erosion control and weed suppression but may interfere with planting operations, maintain soil at temperatures that are too cold in spring, or compete with the crop for limited soil moisture. Management practices that encourage rapid degradation of cover crops such as mowing may reduce effectiveness for weed suppression but may enhance release of nitrogen that can stimulate early crop growth. Soil moisture depletion by cover crops will become the primary management consideration in those areas of the world where soil moisture is the limiting factor in crop production. Cover crop management requires an understanding of all the potential impacts on cropping systems, the definition of the most important goals to be achieved by using cover crops, and a balanced approach for achieving those goals.

Practical Applications

1. Integrate cover crops into a long-term preventive approach to managing weeds that includes a rotational plan to minimize populations as well as appropriate interventions for controlling weeds that do emerge.
2. Rotate cover crops within crop rotations. Continuous use of the same cover crop species or cover crops with the same pattern of planting and growth will select for weed species that are adapted to these species and patterns. Also, cover crops can serve as hosts for nematodes and pathogens and may increase populations of these pests. Cover crops should be rotated in the same way that crops are rotated to reduce buildup of populations of detrimental weeds and pests.
3. Cover crops can permit a reduction of herbicide inputs. Weed suppression provided by cover crop residue usually permits

crops to become established before weeds. Many soil-applied pre-plant or pre-emergence herbicides will be adsorbed to the cover crop residue and become ineffective; use of these products with high levels of cover crop residue may not be economical. However, post-emergence herbicides that are applied to foliage of emerged weeds can be used more effectively with cover crop systems. They may be used only as needed and can be selected for the specific weed species that need to be controlled. This approach could reduce herbicide losses to the environment by replacing pre-emergence herbicides that may be persistent and are often detected in ground and surface waters with post-emergence herbicides that are used at lower rates and are less persistent.

4. Balance management of cover crops for weed suppression with other management requirements. The primary goals of cover crop management may derive from other important benefits of cover crops such as nitrogen contribution to a cash crop or alleviating high soil temperatures. Alternately, the need to minimize negative influences of cover crops such as depletion of soil moisture reserves or interference with planting operations can become important considerations. Successful management of cover crops requires a balanced plan to maximize the benefits and minimize their negatives in order to achieve a productive and sustainable agro-ecosystem.

8

Scientific Process in Culinary Herbs and Aromatic Plants

Consumption Pattern

Americans are consuming ever increasing amounts of fresh, frozen, processed and dried culinary herbs and spices, and this trend appears to be here to stay. The same trend is true for specially fruits and vegetables (Fielding 1988). Factors accounting for increased consumption include interest in new foods and tastes, availability of more fresh herbs, advertising and promotion to food services and institutional food chains, and expanding ethnic populations demanding foods and flavourings of their homeland.

Who will be supplying these herbs and are there opportunities for commercial growers in the production of dried and fresh herbs? Much of the dried herbs produced domestically are produced in California by well-established food companies that own and operate drying and processing facilities and already package, transport and market these products. Growth in this area will continue as long as increased demand and consumption of these herbs continue. Additional growth is likely to occur in the contract growing of herbs by private growers to supply raw product to drying and processing companies.

Some companies both grow and process their own product, but most contract practically all of their raw material needs to growers. Processing companies that are now procuring most of their dried herbs abroad may be more willing to obtain some of the same materials domestically Growth in the dried herbal market will also probably occur with an increasing number of spice companies and natural food wholesalers and retailers catering to specialized markets (*e.g.* organic or pesticide free-herbs and spices, specially restaurants) and which unlike many of the larger food and flavor houses produce specialized herb and other food products. There has been tremendous growth in the fresh herb market as evidenced by the increased variety of herbs available both in the larger supermarkets, the smaller grocery outlets, farmers markets, and roadside farm markets (Simon 1986, Simon et al. 1989, Simon and Clavio 1989, Simon and Grant 1987a, b). Herbs that were only available dried, or fresh during a few months of the year, are now being marketed fresh cut and more recently, as live potted herbs in the produce sections of supermarkets all year long. Recognizing this growth, the USDA (1989c) now lists weekly prices for herbs sold in nineteen major wholesale markets plus the type of container, package, and weight count, and quality of each unit in the National Market News Report. The herbs now covered include anise, arrugala, basil, borage, chervil, chives, cilantro, dill ginger root horseradish, parsley lemongrass, marjoram, mint, oregano, rosemary savoury, sage, sorrel tarragon, thyme, and watercress.

The successful introduction of new culinary herbs into commercial production requires a purposeful strategy and a solid information base. Unfortunately culinary herbs and essential oil crops have been studied little compared to all other food and fiber crops (Craker et al. 1986). While several extension guides have been published on herbs, few guides based on research are available in this country In a bibliographic review of the scientific literature from 1971-1980 (Simon et al. 1984), over 10,000 authors are listed with published research articles on the major economical herbs of the temperate zone involving horticultural research (38%), botany (17%), chemistry (14%), pharmacology (11%), ecology and germplasm (8%). With the exception of peppermint and spearmint most all crop research on herbs was conducted outside of the United States. Although the United States is the principal world producer of mint oils, only 10% of all published mint research from 1971-1980 originated in the U.S. (Craker et al. 1986).

Promising Herbs

There is a wide range of culinary herbs that can be grown in the continental United States and which may offer potential for production. Many herbs are already being commercially produced, albeit in small quantities, often in relatively small farms. Those herbs which show promise for the fresh market are listed. Three examples of culinary herbs with great promise are discussed below.

Coriander (Coriandrum sativum L): This annual herb native to the eastern Mediterranean and southern European region has long been prized for its spicy aromatic seeds which are used either whole or in ground form as a main ingredient in curry and other food and flavor products. Coriander can be grown as a seed crop in the U.S. using existing mechanization and production technology but seeds ripen unevenly and the mature seeds shatter from the plant. Present demand for coriander seed is met by existing foreign suppliers and although the yield of coriander seed is moderately high (1,100-1,700 kg/ha) and the cost of production relatively low, there is still a question whether domestic commercial opportunities exist for coriander seed production as a spice. The profit margin of coriander would at best match the more traditional cash crops (maize, soybeans). Postharvest costs of cleaning the seed, handling and shipping can be major factors determining the profitability of this seed spice crop. Coriander seed is also processed via steam distillation for the extractable essential oil of which d-linalool is the major constituent. The oil is then used in the food and perfume industries.

Recently, American consumers have witnessed the introduction of coriander leaves into the marketplace. This fresh product, known as cilantro, is marketed in bunches as a leafy green spice. While new to American cuisine, cilantro has long been a popular herb in Oriental, Middle Eastern and Latin American cooking.

Coriander, a cool season crop, is easy to grow as a culinary herb and is most suited to fertile loam soils. The plant is direct seeded with a seed drill at rates of 13-18 kg/ha in very early spring. In Florida and New Jersey, coriander is often planted weekly At harvest, the whole plant is manually cut and bunched in the field.

One of the major problems in producing cilantro is premature flowering. Bolting becomes acute as the days get hotter and longer. A number of seed companies now offer slow-to-bolt (long-standing) cultivars. There are significant differences among coriander cultivars regarding the response to premature flowering, and while some are less susceptible, none are totally unresponsive to high temperatures and long days (Simon et al. 1989). Thus, cilantro is planted as a spring, early summer, or fall crop.

The aroma of cilantro is also due to an extractable essential oil, although its composition is distinctly different from the seed oil. Quality of fresh cilantro is based upon strong green colour, strong aroma, and visual appearance (*i.e.* freedom from insects, discolorations). To the grower, yield and the cultivars' resistance to bolt are additional factors to be considered. Cilantro has a relatively short shelf life and requires refrigeration. Cilantro can be kept for 3 to 4 weeks at 0°C, but only 2 to 3 weeks when stored at 5°C (Cantwell 1989). Cilantro is routinely iced after harvest. Cilantro is expected to become increasingly important in the U.S. due in part to the expanding Hispanic and Arabic populations, its unique flavor, and increasing familiarity in the marketplace.

Sweet Fennel (Foeniculum vulgare Mill.): Sweet fennel has highly aromatic leaves and attractive green fern-like foliage but is best known for its seeds which are sold commercially as a spice. Limited domestic production for seed has occurred periodically, but demand is relatively small and adequate supplies can be obtained abroad. Successful domestic production of fennel for the seed as a spice will probably not occur unless the yields can be significantly improved, supplies from abroad become limited, or a government programme makes the production of new crops more beneficial.

Fennel is a perennial but grown as an annual. It is most suited to well drained light loam soil. The plant is easy to grow and is typically direct seeded (3.5-5 kg/ha seed/acre) into the ground in beds with a specialized planter adapted to small-seeded crops. Plant spacing varies significantly with rows in the beds 60-1 00 cm apart and with final plant stands of 10-3 cm apart in the row. Fennel is a cool-season crop, seeds are sown in early spring and germinate at temperatures >7°C (optimum soil temp. approx. 16-18°C). Yields of up to 2,240 kg/ha have been achieved, although earlier USDA estimates were only in the range of 650-900 kg/ha (Sievers 1948). The plant requires 100-115 days to mature before harvest.

The production of fennel as a culinary herb, cultivated both for its aromatic leaves and enlarged leaf stalk is on the rise. This type, called Florence fennel or finocchio fennel, is a different subspecies than regular fennel produced as a seed spice. Very popular in Europe as a specially vegetable in many culinary dishes, it is commonly consumed raw in Italy and cooked in France. Finocchio is becoming commonplace in U.S. supermarkets and consumer interest and familiarity is increasing. One of the limiting factors to increased consumption is the public's unfamiliarity with it's preparation, use and taste. Unfortunately finocchio fennel is often marketed under the misnomer of anise, another culinary herb (Pimpinella anisum L.), which has led to market and consumer confusion. Both plants contain high amounts of anethole in the essential oil, imparting the licorice-like aroma and taste.

The time to harvest fennel bulbs is difficult to assess because the thickened leaf stalk continues to grow and develop until flowering takes place. Care in harvesting, grading and packing fennel must be taken to ensure a high quality fresh pack. Harvesting, cleaning, trimming and packing is done by hand. The foliage must be dark green and fresh in appearance and the stalk and bulb (enlarged base of leaf stalk) a lighter greenish-white colour. The bulb must be firm and the product free from insects and discolorations. Once harvested, fennel should be kept at 0 to 2°C (Seelig 1974). The plants are retailed individually, either wrapped in plastic or simply displayed like celery.

Great variability (in growth and aroma) in commercially available cultivars of fennel and finocchio fennel exist (Simon et al. 1989) making the proper selection of cultivars for spring and fall production important. Finocchio fennel shows excellent potential for future growth in the U.S.

Oregano (Origanum spp.): Their has been a significant increase in the consumption of oregano for both the fresh and processed market (Tucker 1987). Oregano, the common name for a wide number of plant species with a characteristic aroma and flavor, is a perennial aromatic plant native to the dry calcareous soils of southern Europe, southwest Asia, and the eastern Mediterranean (Simon et al. 1984). A major problem which has limited domestic production was in the difficulty in distinguishing the many plant species and types of oregano imported and marketed as oregano. Imported oregano was found to be derived from 16 plant genera and more than 40 plant species (Calpouzos 1954), resulting in oregano being described as a flavor and aroma rather than an individual plant. Taxonomic work by Tucker (1986, 1989) has identified the major types of commercial cultivars and cultivated taxa in this country and the major types of imported oregano. Yet much of the imported oregano arrives as a blend of plants. Selection of an individual line for domestic production, particularly for the dried leaf or as an essential oil, remains difficult.

The European type of oregano comes mainly from subspecies of O. vulgare L. including ssp. hirtum (Link) Ietswaart, ssp. virens (Hoffmanns. & Link) Ietswaart and ssp. viride (Boiss.) Hayek (Tucker 1989). In contrast Mexican oregano, also called Mexican sage, is principally gathered from the small Mexican shrub, Lippia graveolens H.B.K. (Simon et al. 1984), although leaves from other species are collected.

The essential oil of European oregano is composed mainly of carvacrol and thymol. The range of each can be very wide and many chemotypes are available. The herb or the extracted oil is used in a variety of meat and sausage products, salads, stews, sauces and soups. European oregano and to a larger extent Mexican oregano, are used in flavoring Mexican foods, pizza, and barbecue sauces.

Oregano is generally transplanted to the field and grown on light, dry well-drained soils for periods of 3 to 6 years. Domestic horticultural studies on this species are limited. Although yields of more than 14 tonnes fresh herb/ha or almost 4 tonnes dried herb/ha were obtained from one commercial line by a single annual harvest (Simon et al. 1989), typically yields are much lower (1.5-3 tonnes dried herb/ha). Plants can be harvested multiple times each year (from 2 to 6) depending upon the location and end-use. Opportunities exist for dried product as well as for the fresh market. Once harvested for the fresh market oregano should be kept at 0deg.C to maximize shelf-life (Cantwell 1989).

Prospects

Production of culinary herbs for the fresh, frozen processed and dried market will likely increase. The large growth in the production of fresh culinary herbs provides opportunities to growers that either develop a market niche or cooperate closely with a broker or specialist in marketing. It is likely that the large volume herbs will continue to come from areas of intensive vegetable production such as Florida, New Jersey and California.

Herbs can be expected to become fully integrated with fresh market vegetables in packaging, cooling, transport, and marketing operations. Unless governmental regulations change, there will be strong competition from Mexico, the Caribbean, and other areas to supply fresh herbs to the American public. Overproduction of specific herbs within a limited marketplace can result in significant decreases in the wholesale market prices. Greater opportunities for small producers maybe in the development of specialized markets, rather than for the wholesale trade.

Supercritical Fluid Extraction (SFE)

The supercritical fluid extraction (SFE) has been applied only recently to sample preparation on an analytical scale. This technique resembles Soxhlet extraction except that the solvent used is a supercritical fluid, substance above its critical temperature and pressure. This fluid provides a broad range of useful properties. One main advantage of using SFE is the elimination of organic solvents, thus reducing the problems of their storage and disposal in the lipidologist laboratory. Furthermore, several legislative protocols (such as the EPA Pollution Prevention Act in the USA) have focused on advocating a reduction in the use of organic solvents which could be harmful to the environment.

Besides ecological benefits, one of the most interesting properties of SFE is the high diffusion coefficients of lipids in supercritical fluids, far greater than in conventional liquid solvents. Thus, the extraction rates are enhanced and less degradation of solutes occurs.

Several studies have shown that SFE is a replacement method for traditional gravimetric techniques. In addition, carbon dioxide, which is the most adopted supercritical fluid has low cost, is a nonflammable compound and devoid of oxygen, thus protecting lipid samples against any oxidative degradation.

Principles for the Analytical SFE of Lipids

The first guiding principle is the optimization of the solubility of lipids in supercritical CO_2 and the improvement of the fractionation with respect to a particular lipid species.

Important data on the solubility of vegetal oils and other lipids may be found in the detailed study of Stahl et al. (Dense gases for extraction and refining, Springer Verlag, Heidelberg, 1987). These data have led to perform oil and fat extractions above 600 bar and temperatures from 80 to 100°C.

It should be emphasized that many lipid solutes have similar solubility parameters, making their separation by SFE difficult. An improvement in the separation of complex lipid mixture may be found by the integration of adsorbent compound into the extraction cell with the sample.

Several compounds such as alumina, silica, Celite, Florisil or synthetic resins were proposed for the optimization of lipid differential extraction.

The improvement of lipid SFE needs also the study of the kinetics of the lipid removal from the sample matrix. It must be noticed that the fast back-diffusion of analytes in the supercritical fluid reduces the extraction time since the complete extraction step is performed in about 20 min instead of several hours. A common practice in SFE, which must be mentioned in connection with the physicochemical properties of supercritical fluids, is the use of modifiers (co-solvents). These are compounds that are added to the primary fluid to enhance extraction efficiency. Thus, the addition of 1 to 10% of methanol or ethanol to CO2 expands its extraction range to include more polar lipids. When the extraction was performed with supercritical carbon dioxide and 20% of ethanol, more than 80% of the phospholipids were recovered from salmon roe (Tanaka Y et al., J Oleo Sci 2004, 53, 417).

Instrumentation

Those wishing to build their own experimental equipment should read the detailed study of Hawthorne et al. (in Practical supercritical chromatography and extraction, Caude M., Thiebaut D, Eds, Harwood Academic, 1999).

ISCO Inc was one of the first companies to address the off-line SFE market. Extraction cells of 0.5 up to 15 ml are offered.

An automatic version permits automated valve operation for extraction of one or two samples and another model permits 24 samples to be automatically and sequentially extracted. ISCO has also offered a extraction system (FastFat HT) specifically made for the rapid determination of fat content in foods and agriculture products ("The Isco FastFat HT provides rapid, accurate % fat determination in a wide range of foods- a critical parameter in quality control, product labeling, and calibrating on-line process control equipment.

With FastFat HT, you get reliable results in minutes, and eliminate costly delays in production decisions").

Applied Separation offers several extraction units offering great flexibility with respect to sample size and experimental design. Extractor cell sizes can range from several ml to one liter. Off-line and in-line trapping of solutes may be realized.

IECO Corporation is offering a total fat analyzer designed with a triple parallel channel system ("LECO's new TFE2000 Fat/Oil Determinator is specifically designed to determine fat/oil content and eliminate the need for hazardous chemicals used in Soxhlet, Mojonnier/ Roese-Gottlieb, acid hydrolysis, and other solvent extraction methods. Using inexpensive compressed CO2 as the solvent, the analyte is extracted from the sample and transferred to the removable ultra efficient solid phase collection traps. Precise calculations are determined by the integrated balance based on sample weight, and the before and after weights of the collection vials"). Application notes may be found on the LECO site.

The use of high purity SFE-grade CO_2 is not required but impurity and moisture in industrial grade CO_2 can accumulate and may interfere with further analytical operations (gas or liquid chromatography). Thus, an on-line fluid cleanup system may be used to remove trace contaminants.

Extractions can be performed in static, dynamic or recirculating mode. During static extraction, the cell is filled with the supercritical fluid, pressurized and allowed to equilibrate. In the dynalic mode, the fluid is run continuously through the cell, and in the recirculating mode, the same fluid is repeatedly pumped in the cell before being pumped out to the collection vial.

Important Parameters

A. *Sample Matrix:* The levels of lipid content and moisture are important for the analytical process. Lipid levels may be from 1 up to 50% (w/w).

B. *Sample Preparation:* Moisture may inhibit contact between extraction fluid and sample. Thus, removal of water by freeze drying is recommended prior to SFE.

   It may be also efficient to disperse the sample matrix, to grind the sample to increase the mass transfer of lipids. The choice of drying agent can be made by consulting the study of Bulford (Bulford MD et al., J chromatogr A 1993, 657, 413).

C. *Collection of the Lipid Extract:* The collection method for the resultant extract must be optimized to avoid incomplete extraction. The open vials are most frequently used, a sub-ambient cooling being useful for volatile species. It is best to use a collection vial packed with a surface area material, *i.e.* glass beads, glass wool (Snyder J. M. et al., JAOCS 1994, 71, 261) or containing some volume of a chosen solvent. The supercritical CO_2 can be separated from the collected analytes without trouble or disposal problems.

Application of SFE to Lipids

Several reviews were devoted to the application of SFE for the extraction of lipids in various sample types. Those of Clifford A et al. (in Supercritical fluid technology in oil and lipid chemistry, AOCS Press, chap. 19, 1996), Eller FJ et al. (Sem Food Technol 1996, 1, 145) and King J.W. (Grasas y Aceites 2002, 53, 8) should be consulted for specific applications.

Food and various agricultural products were successfully analyzed for their fat and levels by SFE. Progressively, SFE may be regarded as an alternative to organic solvent extraction methods. The extraction efficiencies of SFE and organic solvent methods (Soxhlet or liquid-liquid extractions) were frequently compared and shown to be in good agreement (Eller FJ et al., J Agric Food Chem 1998, 46, 3657). It must be noticed that gravimetric-based results can be influenced by the sample matrix, the moisture and non-lipid moieties and the extraction solvent.

Total fat determinations were done in meat products with two different extractors and compared with a standard method (Berg H et al., J AOAC Int 2002, 85, 1064). The amount and the composition agreed well with results from the standard procedure. To obtain quantitative recoveries by SFE, 1 ml ethanol was added to 1 g sample in the extraction cells before extraction.

Extracting conditions largely affected the composition of dissolved lipids in freeze-dried animal samples (Tanaka Y et al., J Oleo Sci 2003, 52, 295). Thus, one group of triglycerides was extracted from salmon roe while another group remained in the matrix together with phospholipids and a large part of astaxanthin. It is now well known that the extraction yield and composition of triglycrides depend on the extracting conditions. This was verified for the extraction of oil from seaweed (Cheung PCK et al., J Agric Food Chem 1998, 46, 4228), tomato seeds (Bhupesh CR et al., J Food Sci Technol 1996, 31, 137) or soyabean oil (Snyder JM et al., JAOCS 1984, 61, 1851).

The extraction and separation of phospholipids from tuna fish have been described using various concentrations of methanol in supercritical CO2 (Tanaka Y et al., J Oleo Sci 2005, 54, 569). Good recovery of DHA-rich phospholipids were claimed in a series of industrial processes.

A detailed description of a reliable procedure adapted for the determination of total fat in milk- and soy-based infant formula powder has been published by LaCroix DE et al. (J AOAC Int 2003, 86, 86).

The extraction of a specific lipid moiety is generally influenced by substances which are co-extracted, all having a high solubility in supercritical CO2.

One technique to overcome this problem is to use a sorbent to retard the analyte of interest, the interfering substances being removed first. Then, a higher CO2 extraction density or a co-solvent is used to removed the analyte of interest from the sorbent. Thus, it was shown that NH2-bonded silica is able to retard sterols.

Analytical SFE has also been used for lipid-derived volatiles such as aroma in various food samples. For the application of SFE to the study of total fat and lipid classes in meats, the studies of Berg H et al. (J Chromatogr A 1997, 785, 345) and Chandrasekar R. (JAOAC int 2001, 84, 466) should be consulted.

Field of the Invention

The extraction of the principal significant components of herb and spice plants containing pigment, flavor, and aroma and, when present in the starting material, antioxidant, using edible, food-grade solvent.

Of especial interest is the extraction of the principal significant components of spices and herbs represented by plants of the family Solanaceae, representatively Capsicums such as paprika, red pepper, and chili, and Lycopersicon, representatively tomato, all containing carotenoid pigments; Umbelliferae, representatively celery, lovage, dill, carrot, cilantro, fennel, cumin, caraway, parsley, angelica, and anise; Compositae, representatively marigold, artemisia, and tarragon; Leguminosae, representatively fenugreek; Labiatae, representatively rosemary, thyme, sage, oregano, marjoram, mint, savoury, and basil; Zingerberacae representatively ginger, cardamom, and turmeric, Lauraceae, representatively laurel, cinnamon, cassia, and bay; Myrtaceae, representatively allspice and clove; of the genus Myristica, representatively mace and nutmeg; of the genus Piper, representatively black and white pepper; of the genus Vanilla, representatively vanilla; of the genus Allium, representatively onion and garlic; of the genus Sesamum, representatively sesame seed; Cruciferae, representatively mustard and horseradish; the defining characteristic being that it is of a spice or herb plant material from which flavor, aroma, colour, and/or antioxidant can be extracted and used to flavor and/or colour foods and beverages or otherwise employed to enhance the palatability of foods and beverages.

Background of the Invention and Prior Art

The present invention relates to a method of increasing the stability and reducing the microbial counts of both spice or herb oleoresin and the residual cake from which the oleoresin has been extracted. The process simultaneously extracts and concentrates the principal flavor, aroma, colour, and other active components, produces a concentrated and standardized food-grade extract of active components, and a standardized food-grade residual solid, with both the extract and residual solids having significantly reduced microbial counts and frequently also improved stability.

Concentrated extracts of spices and herbs are universally used for flavoring and coloring of food, beverages, and pharmaceuticals. These extracts are traditionally used where a standardized, sterile, and uniform concentrate offers the benefits of control which are inherently difficult to obtain from raw spice or herb, or where the bulk of the raw material is not needed or undesirable.

Ground spice and herb solids are universally used for flavoring, coloring, and imparting otherwise favourable characteristics to food and beverages where the bulk, functional characteristics, and appearance of the food or beverage is important.

Dried spices and herbs, most often in their ground form, are used in the preparation of food and beverages to add flavor, aroma, colour, and preservative properties that make the food more palatable and appealing. The dried spices, ground or unground, are usually added to the food or beverage during preparation at such a point in the preparation that time is allowed for the principal components of interest to be extracted into the food or beverage to impart the desired combination of attributes to the food or beverage.

Further, as spices and herbs are notoriously known to have inconsistent levels of the flavor, aroma, colour, or antioxidants, it is commonly required that spices of varying levels of the principal components of interest be blended to make a final product that is consistent with regard to the principal components of interest to achieve predictable and repeatable performance with respect to the flavor, aroma, colour, or antioxidant release into the food or beverage system in which they are used. This is a costly and time consuming process.

As suggested in prior art, much of the flavor, aroma, and/or colour often is not effectively transferred to the food or beverage. U.S. Pat. No. 2,507,084 overcomes this obstacle of under-utilization of the principal components of interest by first extracting the principal components of interest and subsequently coating the spent spice from the extraction process with a portion of the extract originally removed, thereby extending the useful amount of flavor and aroma that can be derived from a given quantity of spice. It is also disclosed that this process derives value from the exhausted spice solids, from which the flavor, aroma, colour, or antioxidants have been removed, which would otherwise be a waste product. This is a complicated and costly process for recovery of the maximum value of the spice and its principal components of interest.

Traditional extraction processes for the manufacture of concentrated extracts (concentrated several fold as compared with the raw material) involve not only the use of various non-edible solvent systems, but also a large proportion of solvent in relation to the compounds of interest. Many require the use of petroleum distillates, chlorinated solvents, or highly flammable solvents which must be eliminated almost completely from the finished products to make them safe for consumption. These systems require expensive distillation equipment and special precautions must be taken to ensure worker safety and to limit environmental impact.

The intensive processing required often destroys, modifies, or loses some of the more unstable compounds, delicate aromas, flavours, or pigments. More significantly, the last traces of undesirable non-edible solvents are very difficult to separate from the concentrated extract. The residual solid must necessarily contain the same residual non-edible solvents, which are removed only with difficulty. Such residual solvents limit the potential use of the residual solid for human consumption, and are potential environmental contaminants.

Other concentration techniques rely on high pressure equipment to obtain good solvating properties from gases, *e.g.*, liquid or supercritical CO2 (U.S. Pat. No. 4,490,398). High pressure liquefied or supercritical gas extraction requires expensive equipment and has limited solvating abilities for some compounds requiring the addition of cosolvents, or solvents such as propane and butane, which are also difficult to control and may be environmentally sensitive or undesirable in a finished product.

Following extraction and desolventization, the concentrated extract is often standardized with edible solvents and emulsifiers to provide a concentrate with reproducible levels of the active or principal compounds of interest to the user. In an effort to overcome the shortcomings and risks associated with the above-mentioned processes, extraction has been carried out using edible solvents such as vegetable oils or lard. Typical extraction procedures are disclosed in U.S. Pat. Nos. 3,732,111; 2,571,867; and 2,571,948. These methods require a relatively large volume of solvent in relation to the compounds of interest and result in a dilute extract which is limited in its application and which has few of the advantages of the concentrates which can be produced using volatile solvents.

U.S. Pat. No. 4,681,769 discloses a method for simultaneously extracting and concentrating in a series of high pressure countercurrent mechanical presses using relatively small amounts of vegetable oil as the solvent in an attempt to overcome the problem of dilution inherent in earlier processes. This method suffers from severe limitations in temperature and pressure ranges in an attempt to avoid unacceptable oxidative damage, colour loss, yield losses, and flavor changes with the final result being that contact times must be unduly extended for up to 16-24 hours, adding greatly to the cost of the process. Extraction cycle times are unduly long for a given size pressing operation, and the process does not provide for a controlled degree of browning or for sterilization of the extract or of the residual solid. It is also limited to temperatures of less than 100° F. to avoid alleged undue oxidation and thus it does not allow for the use of edible solvents which have a melting point of more than 100° F. or which are highly viscous at temperatures of less than 100° F. Maximum pressures of up to about 500 PSI (cone pressure) are claimed and this severely limits the efficiency and throughput rate for a given size pressing operation, as shown by the disclosure of this patent.

Traditional methods for the sterilization of ground spices and herbs involve the use of extremely toxic substances such as ethylene oxide or methyl bromide, irradiation, or steam and moisture treatment to reduce plate counts to less than 100,000. Chemical sterilization and irradiation of spices and herbs are disagreeable to the consumer because of the perceived risk of residual chemicals and/or radiation remaining in the plant matter and, as a result, several processes using added moisture, such as water or steam, at elevated pressures have been developed as alternatives.

Typical sterilization procedures are disclosed in U.S. Pat. Nos. 4,210,678, 4,790,995, and 4,910,027. All sterilization processes are inherently costly in that they require a separate processing step or steps to accomplish the sterilization, and also present the possibility of further degrading the more unstable components. Addition of moisture or water vapour, as disclosed in U.S. Pat. Nos. 4,210,678 and 4,910,027, prior to or during the heating and sterilization process results in a cooked aroma not typical of the fresh, dehydrated spice or herb and also results in steam distillation and loss of some of the volatile flavor and aroma constituents.

U.S. Pat. Nos. 4,790,995 and 4,910,027 require the addition of a coating of animal protein to protect the spice from the loss of volatile aroma compounds during the sterilization process with water vapour. U.S. Pat. No. 4,210,678 requires bringing the moisture of the spice to above 8-14%, in some cases up to 16-20%, and holding the spice for an extended period of time prior to sterilization to equilibrate the moisture. This additional step is costly and time consuming. In the case of Capsicums, severe browning and off aromas and flavours are developed in the presence of moistures in excess of 10% at elevated temperatures above 180° F.

Traditional methods for controlling the brownness or degree of caramelization of Capsicum solid to enhance its visual appearance involve the use of elevated temperatures and the addition of vegetable or animal fats or oils to bring up the surface colour and luster of the ground spice. This requires a separate and costly processing operation.

Above all, there is the unsolved problem of obtaining satisfactory yields, quality, and throughput rates of acceptable extract having an acceptable content of active principle in the edible solvent without undesirable oxidative damage to, and reduced stability of, the principal compounds of interest, while at the same time providing for simultaneous sterilization of both the herb or spice solid and extract.

Obviously, existing prior art procedures leave much to be desired, and it is a primary objective of the present invention to provide a procedure for the production of sterilized spice and herb products having enhanced stability and which otherwise obviates the shortcomings of the prior art.

Objects of the Invention

Accordingly, it is an object of the present invention to provide a process for simultaneously and rapidly extracting and concentrating the principal components of herb and spice solids, at temperatures of at least 130° F., preferably 130 to about 450° F., in a process which is completely free of petroleum, chlorinated or highly flammable solvent, does not require high pressure gas handling equipment, does not require distillation for solvent removal, uses only food-grade edible solvents which are typically used in the trade to standardize the resulting extract to a desired concentration, and provides a product of reduced bacterial count which is free of adulterants and impurities.

Another object of this invention is to prepare such a concentrated extract by a process which is simple, environmentally friendly, and economical.

A further object of this invention is to prepare a residual solid or press cake which is edible, free of residual petroleum distillates, chlorinated solvent, or other adulterants, which is standardized with respect to the principal components of commercial interest, which has a predictable and controlled degree of brownness or caramelization, and which has a controlled level of water activity with its attendant increased resistance to oxidative deterioration of carotenoid pigments and colour loss where this is otherwise a problem.

Still another object of this invention is to prepare a residual solid or press cake that is standardized with respect to the principal components of interest, that when reground rapidly and efficiently imparts flavor, aroma, colour, and/or other principal components of interest to a food or beverage in which it is used, and all without the use of undesirable non-edible solvents that are inherently difficult to remove from spice solids remaining after extraction prior to standardization with respect to the principal components of interest, are environmentally unfriendly, and are perceived by the consumer as being undesirable in the preparation of a food ingredient.

A still further object of this invention is to provide such a process wherein antioxidants can be added to the edible solvent system so as to protect the concentrated extract and the residual solids against oxidative degradation of the principal components of interest, *i.e.*, flavor, aroma, and colour, which are extracted from the raw plant material or left in the residual solids.

Still a further object of this invention is to prepare an edible extract and an edible residual solid with reduced microbial activity by a process wherein the moisture of the herb or spice is kept below 10%, preferably below 8%, thereby avoiding the loss of volatile flavor and aroma constituents and avoiding the development of uncontrolled browning and off flavor development at temperatures in excess of 130° F. which are necessary to effect high extraction efficiencies, reduction in microbial activity, and improved stability of the carotenoid pigments in both the extract and in the residual solids.

Yet a further object of this invention is to prepare an extract with increased resistance to oxidative degradation of pigments and consequent colour loss. Still an additional object is to provide a process wherein and whereby the residual spice or herb plant solids have their tissue ruptured so as to produce quick release of the flavor, aroma, colour, or antioxidant component therein and thereof when in use in a food or beverage thereby greatly enhancing its use effectiveness, and whereby all products of the process may be conveniently standardized with respect to the principal flavor, aroma, colour, or antioxidant component of interest.

Other objects will be apparent to one skilled in the art to which this invention pertains and still others will become apparent hereinafter as the description proceeds.

Brief Description of the Drawings

The process of the present invention, including the several process steps involved in the simultaneous extraction and concentration of herb or spice solids, *e.g.*, Capsicum solids, to produce the desired extract and sterilized residual solid, both of which have increased resistance to oxidative degradation, and which can be readily standardized to desired levels of the principal components of interest. Although the process illustrated comprises three extraction stages, the number of stages can be decreased to two or increased to more than three to effect the desired relative principal component concentration in the extract and in the residual solid.

Summary of the Invention

The invention, then, inter alia, comprises the following, alone or in combination:

A continuous multistage mixing, high pressure pressing, and countercurrent extraction process for the production of a concentrated edible extract and quick-release edible residual solids, both of reduced bacterial content, and both of which contain herb or spice pigment, flavor, and aroma and, when present in the starting material, antioxidant, from herb or spice plant solids, comprising the following steps: subjecting said herb or spice solids to a countercurrent extraction process involving a plurality of mixing and pressing stages, including first and last mixing stages and first and last pressing stages, together with up to about fifty percent by weight of an edible solvent, to produce an extract and residual solids, continuously returning the extract from each pressing stage to the previous mixing stage, and finally separating the extract from the first pressing stage and separating the residual solids from the last pressing stage, all pressing stages being carried out at a temperature of at least 130° F.; such a process wherein the temperature is 130 to about 450° F.; such a process wherein the solids are subjected to internal pressures in the press stages of at least 6,000 pounds per square inch; such a process wherein the weight of the edible solvent is 5% to about 20% by weight of the solids; such a process wherein the moisture content of the starting solids is less than 10% by weight, and wherein bacterial count reduction is effected at this low moisture content, thereby avoiding undesirable loss of volatile flavor and aroma constituents and avoiding the development of cooked, off flavours and aromas which occur at higher moisture contents; such a process wherein the solids extracted in the process are selected from the group consisting of chipotle, turmeric, black pepper, onion, rosemary, and oregano; such a process wherein the edible solvent is selected from the group consisting of soybean oil, corn oil, cottonseed oil, rapeseed oil, peanut oil, mono-, di-, or triglycerides, lecithin, edible essential oils, sesame oil, edible alcohols, hydrogenated or partially hydrogenated fats or oils, polyoxyethylene sorbitan esters, limonene, edible animal fats or oils, mixtures thereof, and edible derivatives thereof; such a process wherein fine particulate solids are filtered or centrifuged from the extract and alternatively discarded, returned to a mixing or pressing stage of the process, or incorporated in the final residual solids; such a process which includes the steps of hydrating the final extract to add water to the extent of 5% to 200% by weight of the gums and fine particulate solids therein and filtering or centrifuging to remove said gums and solids; such a process including the step of returning the separated hydrated gums and solids to the final residual solids; such a process including the step of rehydrating the final residual solids with water to a water activity greater than 0.3 AW for stabilization thereof; such a process wherein the solids are rehydrated to a water activity of about 0.4 to 0.6 AW; such a process wherein an effective stabilizing amount of an edible antioxidant or chelator is included in the edible solvent; such a process wherein the antioxidant comprises an antioxidant selected from the group consisting of lecithin, ascorbic acid, citric acid, tocopherol, ethoxyquin, BHA, BHT, TBHQ, tea catechins, sesame, and the antioxidant activity from an herb of the Labiatae family; such a process wherein the antioxidant comprises a naturally-occurring antioxidant from an herb of the family Labiatae or powdered ascorbic acid; such a process wherein the antioxidant comprises the antioxidant activity from an herb selected from the group consisting of rosemary, thyme, and sage; such a process wherein the temperature is greater than 180° F.; and such a process wherein the temperature is between about 180° F. and 235° F.

Moreover, an extract of herb or spice plant solids produced by the process having a high level of principal flavor, aroma, colour, or antioxidant components of interest and a low bacterial count due to the high temperature employed in its production and due to the low water content not greater than 10% in the starting solids; and an extract of herb or spice plant solids having a low bacterial count due to the high temperature employed in its production and having improved stability produced according to the process due to edible antioxidant therein; and residual spice or herb plant solid having a low bacterial count due to the high temperature employed in its production and having its tissue ruptured so as to produce quick release of the principal flavor, aroma, colour, or antioxidant component therein, and which is standardized with respect to the principal flavor, aroma, colour, or antioxidant component of interest, produced by the process of the present invention.

The Present Invention

In General

Raw Capsicum or other spice or herb solids, either ground (usually to pass US 40 mesh, and preferably to pass at least US 20 mesh) or unground if coarse particles are desired in the residual solid or cake, *e.g.*, spice solids having a moisture range of about 0.5% to 16% by weight, preferably 0.5 to 12%, and most preferably 1.5% to 10% by weight (ASTA method 2.0), are subjected to a mixing stage, preferably high shear, and in at least one stage an edible solvent is thoroughly dispersed throughout the raw plant material solids.

Typical spice starting plant materials include, for example but without limitation, those of the genus Capsicum including the dried ripe fruits of Capsicum frutescens L. (chilies), Capsicum annum L. (Spanish peppers), Capsicum annum L. var. longum Sendt, its hybrid Louisiana Sport Pepper, and Capsicum chinense (Scotch Bonnet or habenero), all by way of example and not by way of limitation.

Other spices and herbs of interest include those of the family Solanaceae, as stated representatively Capsicums such as paprika, red pepper, and chili, and Lycopersicon, representatively tomato, all containing carotenoid pigments; Umbelliferae, representatively celery, lovage, dill, carrot, cilantro, fennel, cumin, caraway, parsley, angelica, and anise; Compositae, representatively marigold, artemisia, and tarragon; Leguminosae, representatively fenugreek; Labiatae, representatively rosemary, thyme, sage, oregano, marjoram, mint, savoury, and basil; Zingerberacae, representatively ginger, cardamom, and turmeric, Lauraceae, representatively laurel, cinnamon, cassia, and bay; Myrtaceae, representatively allspice and clove; of the genus Myristica, representatively mace and nutmeg; of the genus Piper, representatively black and white pepper; of the genus Vanilla, representatively vanilla; of the genus Allium, representatively onion and garlic; of the genus Sesamum, representatively sesame seed; Cruciferae, representatively mustard and horseradish; the defining characteristic being that it is a spice or herb plant material from which flavor, aroma, colour, and/or antioxidant can be extracted and used to flavor and/or colour foods and beverages or otherwise employed to enhance the palatability of foods and beverages.

The comminuted or uncomminuted plant material is subjected to a plurality of mechanical pressing stages, whereby a concentrated extract of principal components is obtained and a final utilizable and preferably standardized residual solid is produced. The selected edible solvent is introduced into the residual solid at a mixing stage at some point prior to the last pressing stage. The edible solvent, now containing extract, is cycled back to the previous stage, thus always supplying a solvent extract with increasing principal component concentration to the previous mixing and pressing stages. As the extract/edible solvent is passed through each stage countercurrent to the solids flow, a portion of the edible solvent is squeezed or pressed out, thereby extracting a portion of the principal components of interest. As the edible solvent/extract passes countercurrent to the solids, the extracted principal components are progressively concentrated in the extract in a continuous process and the residual principal components end up in the final residual solids known as the cake.

By varying the pressure, temperature, spice solids feed rate, solvent addition rate, and the number of mixing and pressing stages, the concentration of the principal components can be controlled in both the extract and the residual solid.

As will be apparent to one skilled in the art, variations in the process of the present invention can be employed to produce variations in result, the most advantageous of which are the production of both plant material extract of standardized marketable potency and edible residual solid plant material also characterized by standardized marketable potency, and with the edible residual solids also characterized by rapid release of the principal components of interest. Moreover, in the process of the present invention, the spice or herb plant solids have their tissue ruptured so as to provide quick release of the flavor, aroma, colour, or antioxidant component therein when in actual use, *i.e.*, when incorporated into foods or beverages.

For example, using 200 ASTA paprika starting material of about 5% moisture, a 20% soy oil addition, and leaving a residual cake extractable yield of 9.8% by weight of the starting plant solids material, gives an extract with a colour value of 850 ASTA and a residual cake colour value of approximately 50 ASTA. Contrastingly, using a 10% soy oil addition (instead of 20%) yields a cake having about 65 ASTA colour value and, by increasing the residual cake extractable yield to 12.5% by weight of starting plant solids material, the colour value of the residual cake rises to about 100 ASTA and that of the extract to about 1400 ASTA. The lowest colour extract for paprika normally traded is 1,000 ASTA.

Although less than 20% edible oil addition is highly desirable and can be used in many cases, with some edible solvent systems wherein the principal compounds of interest have a limited solubility, or when a more dilute extract and/or lower concentration of principal compounds is desired in the residual solids, more than 20% by weight of edible solvent addition will be required inasmuch as a suitable concentration of principal compounds in the finished extract and in residual solid can in some cases be produced only by the employment of the higher dilution.

Due to the successive treatments of high pressure and pressure relief, with pressures ranging from 6,000 to 30,000 PSI in the pressing stages of the operation, in the presence of added edible solvent, *e.g.*, vegetable oil, and due to frictional heat generated in these high pressure zones, both the residual solid and the extract exiting the process surprisingly have a significantly reduced microbial load over that of the starting material even at moisture levels significantly lower than those indicated by the prior art and, also surprisingly, exhibit increased resistance to oxidative degradation of the carotenoid pigments which are responsible for the characteristic red-yellow colour of various spices, *e.g.*, Capsicums.

The extract from the first or any selected pressing stage may be centrifuged or filtered to provide the finished extract free of particulate solids. Preferably, the fine particulate solids and gums in the extract may be hydrated to about 5% to 200% by weight of the gums and solids prior to centrifugation or filtration to give a crystal clear extract. If water is not used to hydrate the solids and gums, the fine particulate solids from the extract may conveniently be combined with the final residual solids, recycled back into mixing and pressing stages of the process, or alternatively discarded. If water is used to hydrate the solids and gums, it is preferred that the solids and gums be added back to the final residual press solids or discarded.

The edible solvent employed according to the process of the present invention, as illustrated by the following Examples, may be any edible solvent and especially those selected from the group consisting of soybean oil, corn oil, cottonseed oil, rapeseed oil, sesame oil, peanut oil, mon-, di-, and triglycerides, lecithin, essential oils of spices, herbs, or other plants, edible alcohols, propylene glycol, glycerine, hydrogenated or partially hydrogenated fats or oils, limonene, polyoxyethylene sorbitan esters, or any other edible vegetable or animal fat or oil, or mixture thereof, or edible derivatives thereof, the essential aspects of the solvent being that it serves as an extraction aid in which the principal components of the material being extracted are soluble and that it be edible.

The edible solvent, according to the present invention, is combined with the raw material solids to be processed in a proportion of about 5% to about 50% by weight, and frequently amounts as low as 5 to 20% by weight are possible, based on the weight of the starting raw material solids to be extracted. The lower percentages frequently produce a more acceptable and marketable concentration of principal components of interest in both the extract and the residual solids.

The temperature to be employed during the processing and especially in the pressing stages of the process of the invention may be varied widely, but the process is generally carried out at a temperature below about 450° F., and between about 130° F. and 325° F., most preferably above 180° F. and especially between about 180° F. and 235° F.

Temperatures in excess of 130° F. are advantageously employed to achieve acceptable yields and increased throughput rates as compared to the prior art. Higher temperatures are employed to control an increased degree of browning and, most importantly, to reduce the microbial load of both the solids and the extract while at the same time imparting increased resistance to oxidative degradation of the carotenoid pigments of Capsicums in both the extract and the residual solids. Thus, when it is desired that the residual solids from the process have a desirable darkened, caramelized appearance and/ or flavor, a reduced microbial load, and increased resistance to oxidation, this is readily attained by increasing the temperature of the solids and the extract during the process, especially during the pressing stages thereof.

When an antioxidant or chelator is introduced into the process for protection of the spice or herb being processed, this is preferably another plant material or an extract thereof, preferably of the Labiatae family, such as rosemary, thyme, or sage, which is known for its protective antioxidant activity (U.S. Pat. No. 5,209,870), or sesame, or tea catechins, but may alternatively be a suitable edible and preferably an approved food grade additive such as ethoxyquin, BHA, BHT, TBHQ, tocopherol, Vitamin C (*e.g.*, as in U.S. Pat. Nos. 5,290,481, 5,296,249, or 5,314,686), citric acid, EDTA, or the like. The process of the present invention is particularly adaptable to the extraction of any spice or herb plant material solids containing carotenoid pigments or other components which provide colour and/or flavor, pungency, aroma, or antioxidant activity, such as present in rosemary, thyme, sage of the Labiatae genus, to a food with which combined.

Detailed Description of the Invention

The following examples are given to illustrate the present invention but are not to be construed as limiting.

Example 1: Paprika Extraction: Dehydrated paprika (5.5% moisture) is ground in a hammer mill and the resulting ground paprika (95% passing US 40 mesh) is admixed with about 10% by weight of soy bean oil and processed in a countercurrent extraction system involving three (3) pressing stages, each using an Egon Keller Model KEK-100 Screw Press, with the extracts from the second and third stages being returned to the preceding mixing stage before being removed from the process at the end of the first press stage.

A high shear, high speed pin mixer or equivalent is used to mix the soy oil or extracts from the second and third press stages into the ground spice or residual solid from the preceding stage. This recycling is continuous.

The raw material paprika solids are continuously fed at a rate of about 240 lbs. per hour with a total contact time in each mixing stage of about 15-60 seconds. The residence time in each press is 5-60 seconds. The pressing stages are operated at about 10,000 PSI internal pressure and about 200 degrees Fahrenheit, which is maintained by cooling with water through the bore of the press shafts.

The starting colour value of the ground paprika solids is 200 ASTA. The principal components extracted and standardized in both the extract and the residual solid are the carotenoid pigments. The resulting final soy-paprika extract has a colour value of about 1,375 ASTA and the reground paprika residual solid from the final (3rd) press stage has a colour value of about 85 ASTA.

Example 1A: Variation: By varying the percentage of edible solvent employed from about 5% to 20%, the pressure from about 6,000 to 30,000 PSI, the number of countercurrent mixing and pressing stages from 2 to 5, with return of the extract from each press stage to the preceding mix stage before final removal from the process in the first press stage, varying the temperature from about 130° F. to 280° F., and removing the seed from the paprika solids prior to grinding, the resulting extract ranges in colour value from about 2,700 ASTA to about 800 ASTA and the residual solids range in colour value from 180 ASTA to 35 ASTA.

By regrinding the residual solids (from the final stage) just as is done with fresh, dehydrated paprika, a product in every way comparable to commercially available ground paprika solids is produced. After filtering or centrifuging off the fine particulate solids, the extract can be directly substituted for commercially available paprika oleoresin in every respect.

By varying the pressing temperature of the process from about 130° F. to 325° F., the hue of the reground residual solid is varied from slightly browned to a dark chocolate brown, demonstrating that the degree of brownness can be controlled by the pressing temperature employed. The degree of "brownness" is measured using a Hunter Labscan Spectrocolorimeter with 0 degree illumination, 45 degree circumferential viewing, illuminant D65, 10 degree observer, Ceilab coordinate system.

The hue of the paprika powder is measured by placing the powder in a 2.5-inch diameter cuvette, shaking gently to ensure even coverage, and measuring through the bottom of the cuvette. The results of the varied operating temperatures of the process are shown in Table. The designation L is indicative of the "lightness" of the sample with the higher numbers being lighter or less browned, and the lower numbers being darker or more browned.

Processing Temperature	*Visual Appearance*	*L* Values*
130° F.	Red	40.18
150° F.	Tan-Red	37.25
200° F.	Light Brown Red	33.22
280° F.	Dark Brown Red	29.16
325° F.	Chocolate Red	22.85

The data clearly demonstrate that the degree of browning can be controlled by varying the press temperature at which the process is conducted. This broadens the applications or uses of the residual solid to include a base for toasted chili powder and as a replacement for browned, caramelized paprika. The residual solid can be substituted for ground paprika or chili powder in many common applications and a separate processing step for browning to a desired degree is not required.

The starting ground paprika solids have an aerobic plate count (Analysis run according to Bacterial Analytical Manual By AOAC, 8th edition, 1995, and ISO-GRID Methods Manual, 3rd edition, 1989) of about 14,000,000. The residual solids exiting the extraction system have a count of about 2,000 to 200,000, with the lower count being achieved at the higher temperatures. This is a significant reduction and makes the residual solids per se suitable for any application where treatment with ethylene oxide or irradiation would normally be required.

Example 1B: Antioxidant Addition: The foregoing example is repeated with all materials and conditions being the same, except that the soybean oil edible solvent is supplemented with an antioxidant blend at a concentration of 3% by weight of the original ground paprika solids. The blend consists of about 29% lecithin, 20% powdered ascorbic acid, 5% citric acid, 15% tocopherol, and 1% rosemary extract (in accordance with Chang and Wu U.S. Pat. No. 5,077,069).

The stability of (1) the resulting extract and (2) the residual solids is compared in each case with an untreated control. In such evaluation, the paprika extracts are plated on flour salt to an extent of 2.4% by weight with a mortar and pestle. Two-gram samples are weighed into 13×100 mm test tubes. The test tubes are stored in a thermostatically-controlled oven at 65° C. Samples are withdrawn periodically, extracted with acetone, and the colour at 460 nm of a standard (%) dilution in acetone is determined spectrophotometrically.

In the evaluation of the residual solids, two-gram samples of the reground residual solid are substituted for the flour salt dispersions.

The procedure for the "standard dilution" is as follows: The initial colour of the dispersion is determined by pouring two grams of the original dispersion into a 100-ml flask. Acetone is added up to the 100-ml level.

The flask is inverted several times. The flour salt is allowed to settle for five minutes. Then three ml of the dilution is pipetted into a 25-ml flask and diluted up to the 25-ml level. The absorbance is read at 460 nm. The 460 nm colour is determined by the formula: ##EQU1## to translate to ASTA colour, multiply the 460 nm colour by 820.

The colour is plotted against time and the time for 1/3 of the starting colour to fade is reported as the 2/3 life.

This is a highly-reproducible measurement, which is sufficiently accurate to evaluate the effectiveness of the antioxidants and will assist the practitioner to optimize formulations for specific uses.

The final extract from the first press stage of the unprotected or unstabilized process has a colour value of about 1375 ASTA and a 2/3 life of 6.5 hours as compared to a colour value of about 1600 ASTA and a 2/3 life of 63 hours for the extract from the protected material. The colour value of the unprotected or unstabilized residual solids is about 85 ASTA with a 2/3 life of 54 hours, compared to the protected solids which have a colour value of about 95 ASTA and a 2/3 life of 155 hours. This clearly demonstrates that inclusion of antioxidants can improve not only the colour yields from the extraction process but also at the same time improve the colour stability of both the extract and the residual solids.

Other suitable antioxidants (*e.g.*, lecithin, ethoxyquin, butylated hydroxy anisole (BHA), butylated hydroxy toluene (BHT), tertiary butyl hydroxy quinone (TBHQ), sesame, tea catechins, and Labiatae herb antioxidant activity, finely-divided ascorbic acid, tocopherol, citric acid) can be substituted in whole or in part for the specific antioxidant mixture employed with similar desirable colour-protective results, preferably a naturally-occurring antioxidant from an herb of Labiatae family, *e.g.*, rosemary, sage, or thyme, or powdered ascorbic acid.

Example 2: Effect of Varying Operating Temperatures: Dehydrated paprika solids (2.5% moisture) were ground in a hammer mill and the resulting ground paprika (95% passing US 40 mesh) was processed with about 15% by weight of soy bean oil in a countercurrent extraction system as in Example 1 involving two (2) pressing stages, with extracts from the second press stage being returned to the preceding (first) mix stage before being removed from the process at the first press stage.

Upon exiting the first press stage, distilled water was metered continuously into the crude extract at a rate of 75% by weight of the gums and solids by means of an inline static mixer. The weight of the gums and fine particulate solids in the extract was determined by diluting one gram of the crude extract in nine grams of acetone. The mixture was spun down for three minutes at 2000 G's in a laboratory centrifuge. The solids separated were air dried and the weight of the gums and solids was calculated as a percentage of the weight of the starting extract. The hydrated gums and solids removed from the extract were continuously returned to the final residual press solids via a high shear, continuous pin mixer installed immediately following a water-jacketed cooling screw which received the residual solids from the second press stage.

Prior to hydration and centrifugation, the extract contained approximately 10% by weight of gums and fine particulate solids as determined by the above-described method. Following hydration and centrifugation the gums and particulate solids amounted to no more than 1% by weight of the extract and the extract was a crystal clear solution, free of any suspended insoluble materials.

The colour value of the starting ground paprika was about 150 ASTA. The pressing stages were operated at about 20,000 to 30,000 PSI. The extraction process was started with the presses operating at about 80° F. as measured by the temperature of the cake exiting the presses.

The temperature of the presses was controlled by the rate of flow of cooling water through the bore of the press shafts and the screen cages to keep the operating temperatures in the range of 80° to 180° F.

Over the time of the extraction run, the operating temperatures of the presses, as measured by the temperature of the cake exiting the presses, was gradually increased to about 255° F. by first slowing and then stopping the flow of cooling water to obtain operating temperatures of 180-200° F., and then by substituting steam for the water in the shaft and cages at gradually increasing pressures to achieve temperatures of 200-255° F.

Samples of the extracted oil and press residual solids were pulled at various temperature intervals as the temperatures were increased. Samples of the residual solids were pulled at two points, the first (non-rehydrated) immediately after exiting the cake-cooling screw following the final (second) pressing stage, and the second after the thus-cooled residual press solids were rehydrated to a moisture content of about 10%.

The samples were assayed for ASTA colour, aerobic and anaerobic plate count, and colour stability over time using methods employed in Examples 1A and 1B.

The advantages of operating the process at a temperature above 130° F., as indicated by the temperature of the cake exiting the presses, can clearly be seen. The plate count of both the extract and the cake are progressively reduced as the temperatures are increased.

Table: *Effect of Increasing Temperatures on the Plate Count of the Extract*

Temperature Degree F	*Aerobic Plate Count*	*Anaerobic Plate Count*
80	1,900,000	790,000
130	1,700,000	800,000
150	1,700,000	660,000
170	1,600,000	590,000
175	1,500,000	425,000
180	1,300,000	380,000
190	360,000	150,000
200	300,000	200,000
215	240,000	150,000
225	190,000	65,000
235	170,000	32,000
245	69,000	8,600
255	3,800	830

Table: *Effect of Increasing Temperatures on the Plate Count of the Press Solids*

Temperature Degree F	Aerobic Plate Count	Anaerobic Plate Count
80	220,000	55,000
130	160,000	35,000
150	160,000	25,000
170	100,000	20,000
175	32,000	15,000
180	80,000	7,400
190	3,500	800
200	9,800	3,400
215	5,800	2,300
225	4,100	500
235	1,900	1,100
245	5,400	100
255	800	100

The efficiency of extraction is dramatically improved as evidenced by the progressively decreasing ASTA values and the progressively decreasing residual extractable yields of the press residual solids. It is apparent that, to achieve residual extractable yields of less than about 20% by weight of the cake, it is necessary to operate the presses at 130° F. or higher. Moreover, for obvious reasons of efficiency, temperatures above 180° F., and especially between about 180° F. and about 235° F., are greatly preferred.

Table: *Press Cake ASTA and Residual Yields at Progressively Increasing Temperatures*

Temperature Degree F	Press Solids ASTA	Press Solids Residual Yield
80	87	28.28%
130	76	16.40%
150	65	15.72%
170	61	15.72%
175	53	12.36%
180	43	13.88%

Contd...

Temperature Degree F	*Press Solids ASTA*	*Press Solids Residual Yield*
190	42	10.84%
200	44	10.72%
215	41	9.96%
225	39	9.50%
235	33	9.28%
245	32	9.00%
255	35	9.80%

Most importantly, the stability of the extract is not adversely affected and is in fact increased. The accelerated study was done according to the procedures described in Example 1B with the colours reported as a percent of the starting colour for each respective sample to adjust for the varying colour yields at the respective temperatures. These results demonstrate that the extract produced at higher operating temperatures exhibits increased resistance to oxidative colour deterioration. This is surprising, as explained in the following.

Table: *Press Oleoresin (Extract) Stability, Accelerated, 65° C.*

Temperature Degree F	*Hour 2*	*Hour 4*	*Hour 8*	*Hour 12*	*Hour 17*
80	94%	88%	81%	73%	62%
130	94%	89%	82%	75%	64%
170	93%	89%	82%	76%	65%
225	94%	90%	82%	78%	67%
235	94%	90%	82%	77%	69%
255	95%	90%	84%	78%	72%

It is commonly believed that lipid-containing systems, when exposed to heat, will exhibit an increased rate of lipid oxidation that, once initiated, will proceed at an ever-increasing rate. (Rancidity and its Measurement in Edible Oils and Snack Foods, A Review, Robards, Kerr, and Patsalides, Analyst, February 1988, Vol 113). In fact, prior art (U.S. Pat. No. 4,681,769) claims a process for counter-current, high pressure extraction of Capsicums at less than 100° F. and less than 500 PSI for the express reason of protecting the extracted oil from oxidation.

To confirm the positive effect of high temperature treatment in more controlled conditions, a forty gram sample of hexane-extracted oleoresin paprika, with no diluents added, was heated in a beaker on a heated stir plate at 100° C. for eight and one-half hours.

A control sample which was unheated, a sample pulled from the heated beaker after four hours, and a sample of the material heated for the full eight and one-half hours were dispersed on flour salt to make dispersions of 1.2% oleoresin by weight of flour salt. Two gram-portions of the dispersions were weighed into test tubes and placed in a 65° C. oven. An initial ASTA colour was run on each dispersion and then ASTA colours were run periodically and the results were plotted versus time to determine the relative stability of the heated and unheated samples.

Table: *ASTA Values of Heated & Unheated Oleoresin Paprika (Extract) Over Time*

Heated 4		***Heated 8***	
Hours	Unheated	@ 100° C.	@ 100° C.
0	32.5	31.5	26.0
2	29.0	29.0	25.8
4	26.0	28.0	25.7
6	24.0	27.0	25.5
8	22.5	25.8	25.3
10	21.0	24.5	25.0
12	20.0	23.0	24.8
14	19.0	22.3	24.5
16	18.0	21.8	24.0
18	17.0	21.0	23.5
20	16.0	20.0	23.0
22	15.1	19.0	22.5
24	14.2	18.5	22.1
26	13.4	18.0	21.8
28	12.9	17.5	21.4
30	12.5	17.0	21.0

The non-rehydrated press residual solids produced in Example 2 exhibit decreased resistance to oxidative colour loss at the press operating temperatures are increased as predicted by prior art (Bennett et al, U.S. Pat. No. 4,681,769).

Table: *Stability of Non-Rehydrated Press Solids at Various Press Operating Temperatures, Expressed as Percent of Starting Colour Retained*

Temperature	***Week***	***Week***	***Week***
Degree F	2	4	6
80	86.7%	82.2%	85.5%
130	89.6%	85.5%	84.6%
170	73.3%	65.3%	58.1%
225	61.7%	35.8%	32.5%
245	68.2%	31.0%	19.3%

But, very importantly, it can be seen that the press residual solids which are rehydrated immediately after exiting the second press stage of the process (Example 2) exhibit significantly increased stability relative to the non-rehydrated solids, thus overcoming the claimed disadvantages from operating at temperatures above 100° F. as set forth in U.S. Pat. No. 4,681,769.

Table: *Stability of Rehydrated Press Solids at Various Press Operating Temperatures, Expressed as Percent of Starting Colour Retained*

Temperature	***Week***	***Week***	***Week***
Degree F	2	4	6
80	90%	92%	91%
130	93%	91%	92%
170	92%	92%	91%
225	94%	93%	91%
245	95%	94%	93%

In fact, after discounting for the effect on pigment stability of increasing residual extractable yields in the press solids obtained at the lower temperatures, the carotenoid pigments in the residual solids would show enhanced stability for a given residual extractable yield. These are surprising and unexpected results and clearly overcome the supposed obstacle of operating at elevated press temperatures and pressures.

It is further surprising that the colour stability of the residual press solids is significantly improved by controlling the water activity

(AW) of the solids in ranges above those suggested for the stabilization of lipid-containing systems by extensive studies and particularly by Nelson and Labuza, Water Activity and Food Polymer Science: Implications of State on Arrhenius and WLF Models in Predictina Shelf Life, K. A. Nelson & T. P. Labuza, Journal of Food Engineering 22, 271-289 (1994). Water activity is defined as the ratio of the vapour pressure of water in a food to the vapour pressure of pure water at the same temperature. Prior art suggests that maximum stability of lipid systems should be attained at water activities of about 0.3 with decreasing stability developing as the water activity is increased above this level. In this example we find precisely the inverse effect on stability of the carotenoid pigments for a given water activity.

In order to confirm the effect of high temperatures in the pressing operation, and to confirm the effect of added moisture, a controlled test was performed on a laboratory scale where the effect of levels of extractable yield in the cake could be controlled to eliminate the effect of variable press cake residual yields on the stability of the carotenoids. A 3,000 gram sample of ground paprika solids (175 ASTA, 9.8% extractable yield) was dried in a lab tray dryer at 100° F. for 16 hours to a moisture content of about 2%. One half of this sample was then heated in an oven at 220° F. for twenty minutes to approximate the temperature in a pressing operation according to the invention.

The other unheated sample served as a control. One hundred gram samples of each of the two materials were rehydrated at approximately 1% intervals up to about 12% moisture. The water activity AW of each was determined using a Rotronics Hygroskop DT, model DT2/1-00IV, water activity instrument. Samples were weighed into sealed test tubes, stored at ambient temperatures of about 72° F. in the dark, and the ASTA colours were determined over a period of eighteen weeks to determine the relative rates of colour degradation. The colour retained (as a percentage of the starting colour for each sample to compensate for the effect of colour dilution with the rehydration water) was plotted against time.

Table: *Percent Colour Retained of Unheated Ground Paprika at Various Water Activity Ranges*

Water Activity Aw	*Week 1*	*Week 5*	*Week 18*
0.15	74%	57%	42%
0.30	50%	45%	12%

Contd...

Water Activity Aw	***Week 1***	***Week 5***	***Week 18***
0.40	68%	50%	43%
0.60	83%	68%	55%

Table: *Percent Colour Retained of Heated Ground Paprika at Various Water Activity Ranges*

Water Activity Aw	***Week 1***	***Week 5***	***Week 18***
0.15	66%	56%	41%
0.30	60%	50%	45%
0.40	80%	62%	57%
0.60	98%	82%	78%

It can be seen in Tables 9 & 10 that the stability of the carotenoid pigments follows almost precisely the inverse of the curve predicted by Nelson & Labuza. It can also be seen from these tables that controlled temperature (with concurrent browning) significantly enhances the stability of the carotenoids above a water activity of 0.3 and particularly in the water activity range of 0.4 to 0.6. Water activity ranges higher than 0.6 were not tested as levels marginally higher than this range will support microbial growth which is not acceptable in a dry spice product.

It can be concluded that the stability of the carotenoid pigments found in Capsicums unpredictably does not follow the commonly-accepted and predicted pattern for lipid oxidation with respect to temperature and water activity as suggested in U.S. Pat. No. 4,681,769, or in the cited literature (Nelson and Labuza, Water Activity and Food Polymer Science: Implications of State on Arrhenius and WLF Models in Predicting Shelf Life, K. A. Nelson & T. P. Labuza, Journal of Food Engineering 22, 271-289 (1994); Rancidity and its Measurement in Edible Oils and Snack Foods, A Review, Robards, Kerr, and Patsalides, Analyst, February 1988, Vol 113); describing the stability of lipid systems.

In fact, high temperature treatment, combined with rehydration of the press solids to a water activity above 0.3, preferably of 0.4 to 0.6, significantly improves stability rather than decreases it. This is a very surprising and unpredicted result.

It is well known that the lipid profile of Capsicum and its extracts, without the addition of any diluents, comprises a mixture of saturated and unsaturated fatty acids, 60-70% being unsaturated linoleic and linolenic, Lipid and Antioxidant Content of Red Pepper, Daood, Biacs, et al., Central Food Research Institute, Budapest, Hungary (1989) and The Nature of Fatty Acids and Capsanthin Esters in Paprika, Nawar et al., Journal of Food Science, Vol 36 (1971).

In fact, Daood et al suggest that "...the presence of triglycerides containing high amounts of unsaturated fatty acids may be an important factor contributing to the fading of paprika during processing and storage." The present findings are just the opposite. Without in any way being limited by theoretical considerations, it is hypothesized that the presently-discovered surprising and unpredicted inverse relationship shown (in Tables 9 & 10) between the stability of carotenoid pigments at given water activities is due to the fatty acids in the substrate being Preferentially attacked by the oxidation reaction at the low (from about 0.05 to 0.2 AW) and higher water activity ranges (above 0.3, preferably about 0.4 to 0.6 AW), thus protecting the carotenoids. At the intermediate water activity ranges (0.2 to 0.4 AW), where the lipids are best protected, the carotenoids are more readily and preferentially attacked and exhibit low resistance to oxidative degradation.

Another controlled test was conducted to demonstrate the effect of different extractable yields in the residual solid press cake. The effect of higher amounts of unsaturated fatty acids is evident from the results illustrated where fresh, refined, bleached, and deodorized soybean oil with no antioxidants was added at various percentages based on the weight of the paprika. The colour over time was compared to the untreated control in an accelerated study at 65° C.

A typical Refined, Bleached, and Deodorized soy oil has a fatty acid composition of 22.3% Oleic (18:1), 51% linoleic (18:2), and 6.8% linolenic (18:3). (Riegel's Handbook of Industrial Chemistry, 9th Edition, pg 278). It can be concluded that higher levels of unsaturated fatty acids, such as oleic, linolenic, and linoleic, which are found in most vegetable oils, will improve the colour stability of the press residual solids.

Levels of extractable yield in the residual solids above about 15-20% by weight of the residual solids is undesirable as the residual Capsicum solids become difficult to handle for most uses and the efficiency of extraction is reduced, *i.e.*, less colour can be removed from the spice as the residual yield is allowed to increase by decreasing either the pressure or temperature employed.

Table: *Percent Colour Retained with Varying Amounts of Soy Oil Added to Ground Paprika*

Percent Addition	*Hour 2*	*Hour 4*	*Hour 6*	*Hour 8*
0%	65%	59%	52%	50%
5%	90%	83%	74%	72%
10%	2%	84%	75%	74%
15%	94%	87%	80%	78%
20%	96%	91%	83%	81%

It is readily apparent, comparing the results of the controlled test (Tables 9 & 10) on stability of heated vs unheated material, where oil is controlled at a constant level that, at a given added soy oil content in the press residual solids, the colour stability of the residual press solids is significantly improved when the spice, *e.g.*, a Capsicum, has been exposed to higher temperatures.

This conclusion is not readily apparent in the results shown in Table where the amount of residual vegetable oil left in the press residual solids is higher in the low temperature ranges due to the decreased efficiency of the extraction process at lower temperatures. The presence of higher amounts of residual oils there offers some protection which overshadows the increased protective effect at higher temperatures so evident in Tables.

It can therefore be concluded that much, if not all, of the protection offered by operating the presses at temperatures lower than 100° F. (as claimed in U.S. Pat. No. 4,681,769) as compared to temperatures above 100° F. is simply due to the higher residual oil levels (reduced extraction efficiency) and that, for any given residual oil content, and with rehydrated residual solids, the operating temperatures above 130° F. give superior results, not only in an increased extraction efficiency which allows for a continuous, high speed process with increased throughput rates and significantly reduced microbial activity, but most surprisingly in an increased colour stability of both the extract and the residual press solids, particularly when the press solids are rehydrated.

Comparative Example: According to Bennett U.S. Pat. No. 4,681,769,—Low Temperature and Pressure.

The press solids residual yield is much higher at temperatures below 100° F. and much higher (28.3% residual yield) than disclosed in U.S. Pat. No. 4,681,769 (10-15% residual yield). In an effort to more closely model the residual yields of 10-15% (oil) in the cake as disclosed in U.S. Pat. No. 4,681,769, the feed rate for this test was set at about 95 pounds per hour, thus allowing more residence time in the press to expel more extract and to reduce the residual yield of the press residual solids to 10-15%.

Various Non-capsicum Spice or Herb Plant Solids

In addition to the foregoing, the following examples show the applicability of the process of the present invention in the production of herb or spice plant extracts and solids having greatly reduced bacterial counts and without the deterioration alleged by the prior art to occur at temperatures above about 100° F., especially at low water contents which are also contraindicated by the prior art, and with the production in each case of residual spice or herb plant solids having its tissue ruptured so as to produce quick release of the flavor, aroma, colour, or antioxidant component therein when in actual use in a food or beverage, thereby greatly increasing the use effectiveness thereof, and both extract and solids produced by the process being readily standardized with respect to the principal flavor, aroma, colour, or antioxidant component of interest.

Example 3: Paprika Extraction and Flavor/Aroma Activation : Paprika is processed in the same manner as given in Examples 1, 1a, 1b, or 2 using a multiple stage, counter-current process with temperatures being in excess of 130° F. with the residual solids being rehydrated to a water activity of 0.6. By adjusting the amount of edible, food grade solvent, and the pressures and temperatures in the pressing stages, the residual cake solids are standardized with respect to ASTA colour value to produce paprika solids with ASTA colour values equivalent to ground paprika commonly used commercially as described in the above examples.

A sample of the residual cake solids at 120 ASTA was reground to pass US 40 mesh and was compared to the starting, unextracted spice, also ground to pass US 40 mesh and rehydrated to a water activity of 0.6, with respect to flavor and aroma intensity. Ten grams each of the residual cake solids and the unextracted spice were thoroughly mixed with 50 grams of flour salt to make a dispersion that would typify a coating for snack foods used to impart a paprika flavor and aroma. Triangle tests were done on the samples.

The samples were coded with random three digit numbers and twelve expert panelists were asked to identify the sample that had the highest paprika flavor and aroma. Twelve of twelve panelists picked the reground residual cake solids as having a more intense paprika aroma. Eight of twelve picked the reground residual cake solids as having a more intense paprika flavor. By statistical analysis (according to methods described in Sensory Evaluation Techniques, Vol II, by Morten Meilgaard, 1987, page 133) this clearly demonstrates that the residual cake solids have more intense paprika aroma and flavor available in a food system as a result of the high pressure, high shear forces and the attendant disruption of the flavor and aroma containing tissue of the spice.

Example 3a: Colour Activation : A sample of 120 ASTA ground residual cake solids was prepared as in Example 3 above. This was compared to a sample of commercially available 120 ASTA paprika with regard to effective colour release into water to model colour release under conditions typical of those used to prepare a marinade or sauce. One (1) gram each of the ground residual cake solids and commercial ground paprika were added to 100 ml of distilled water and stirred for two minutes at ambient temperature.

The slurries were then filtered to remove all particulate solids and the absorbance of each of the solutions was measured spectrophotometrically at 460 nm. The absorbance of the water solution prepared using the residual press solids was 0.566. The absorbance of water solution prepared with the commercial ground paprika was 0.294. This clearly demonstrates that the effective utilization of the colour, and the colour intensity of the model sauce system, were significantly enhanced by the high pressure, high shear, and high temperature of the pressing and extraction process, which ruptures the tissues of the plant solids.

Example 4: Chipotle Extraction and Flavor/Aroma Activation : Chipotle (smoked and dehydrated Jalapeno capsicum pepper) at 4% moisture is ground to pass US 40 mesh and extracted in a two-stage countercurrent pressing operation with the addition of 6% of an edible solvent mixture consisting of 95% soybean oil and 5% decaglycerol tetraoleate (Mazol PGO 104K) by weight of the starting spice in the same manner as described in the above Examples. The principal components of interest in this spice are the flavor (including capsaicinoids which impart the pungent flavor typical of chili peppers) and aroma compounds (including the smokey aroma).

The pungency of the starting ground Chipotle is 35,000 Scoville (ASTA method 21.0). The pungency of the extract is 200,000 Scoville and the pungency of the residual cake solids is 20,000. Both the extract and the residual cake solids have the characteristic smokey aroma of the starting spice. By varying the addition rate of the soybean oil, the concentration of pungent principles is varied in both the extract and the residual solids. For example, using a 12.5% soy oil addition results in an extract with a Scoville pungency of 165,000 and a residual cake solid of 15,000 Scoville. Increasing the soy oil addition to 15% results in an extract of 150,000 Scoville and a residual solid of 7,500 Scoville. This clearly demonstrates that the spice solids can be predictably standardized effectively and efficiently with respect to the principal flavor compounds without the requirement of traditional blending and replating of solvent-extracted oleoresins as described in prior art (U.S. Pat. No. 2,507,084).

The aroma of the residual cake solids is compared to that of the starting unextracted spice using the Triangle method described in Example 3 by dispersing 10 grams of each in 50 grams of flour salt. Six of the twelve panelists can not pick the extracted Chipotle residual solids as being different from the unextracted starting spice, clearly demonstrating that the high pressure, high shear process provides for a ruptured spice residual solid with equivalent aroma intensity even though a portion of the principal components of interest have been removed with the extract.

The starting raw spice has an aerobic plate count of 80,000,000. The plate count of the final extract and residual cake solids are 15,000 and 1,100,000 respectively, again demonstrating that significant reduction in microbial activity can be achieved at such low water contents.

Example 5: Onion Extraction—Flavor and Aroma Activation : Dehydrated, toasted onion powder (Gilroy Foods, Inc. 92700 Toasted Special Onion Powder) was processed in a high pressure extraction system comprising two pressing stages, using Egon Keller Model KEK-100 Screw Presses. Glycerol monooleate containing 10% propylene glycol (Mazol DF300K) was added to the high shear mixing stage prior to the second pressing stage at 25% by weight of the starting spice. The raw material onion was continuously fed at a rate of about 150 lbs per hour with a total contact time in the pressing stages of about 120 seconds. The pressing stage was operated at about 6,000 psi internal pressure and about 190° F., which was maintained by cooling with water through the bore of the press shaft as needed.

The principal components of interest in both the extract and the residual solids are flavor and aroma constituents typical of toasted onion powder. The resulting final extract from the pressing process has the characteristic aroma and flavor profile at a level approximately 4 times the strength of the starting spice as determined by plating the extract on flour salt at varying concentrations and subjecting the resulting dispersions to triangle tests as described in previous examples compared to the unextracted spice. This clearly demonstrates that the flavor and aroma components of the raw spice can be effectively extracted using high pressure, high temperature mechanical pressing.

The residual cake solids are compared to the unextracted spice by dispersing 20 grams of each in 200 ml of salad oil as in Example 3a above, stirring for 5 minutes, and filtering to remove all particulate solids to compare the effectiveness of flavor release into a model salad dressing system. Twelve trained panelists are asked to pick which salad oil has a more intense toasted onion flavor. Twelve of twelve pick the salad oil prepared with the residual cake solids as having superior and more intense toasted onion flavor and aroma. This is a very surprising and unexpected result as a significant portion of the flavor and aroma of the starting spice was removed with the extract.

Without in anyway being limited by theoretical considerations, it is hypothesized that the nature of the edible solvent system used, in combination with the high temperature, high pressure, and high shear forces generated during the processing of the spice, result in a ruptured spice tissue matrix with the flavor and aroma compounds remaining in the residual cake solids being readily soluble in salad oil which otherwise does not readily dissolve the principal components of interest. It is clear that this process provides very significant economic and functional value over prior art.

Example 6: Black Pepper Extraction, Standardization, and Flavor Activation : Black Pepper, ground to pass US 30 mesh, was subjected to a two-stage, countercurrent extraction process as demonstrated in the previous Examples using Triacetin (glycerol triacetate) as the edible solvent at 7% by weight of the starting spice. The temperature of the residual cake solids exiting the pressing stages was maintained at about 215° F.

The piperine content of the starting spice was 11% by ASTA method 12.1. Piperine is responsible for the characteristic pungent flavor of black pepper. The final extract, after filtering to remove the fine particulate solids, had a piperine content of 41.6% and the residual cake solids had a piperine content of 4.1%, typical of commercially-traded grades of black pepper commonly used in food preparation.

An expert panel compared the ground residual cake solids to commercial ground black pepper at 4% piperine. The panelists unanimously picked the residual cake solids as being more pungent, demonstrating that the high pressure, high shear treatment in combination with an edible food solvent provide a surprisingly improved residual spice solid, as compared to an untreated spice having equivalent levels of the principal compound of interest. This presents a clear advantage over prior art.

Example 7: Rosemary Extraction and Standardization of Antioxidants: Ground, dried Rosemary at 6% moisture was processed in accordance with the procedures described in Example 6 but using a medium chain triglyceride based on caprylic/capric fatty acids (Neobee M-5, produced by the Stepan Company) at a level of 40% by weight of the starting spice, with temperatures of the residual cake solids maintained at about 250° F. with cooling water or steam as needed through the bore of the press shafts.

The starting raw material had levels of Carnosol and Carnosic acid, two of the primary antioxidant compounds present in rosemary, of 0.46% and 2.47% respectively, for a total of 2.93% by weight of active compounds. Carnosic acid and Carnosol levels were determined by extracting a sample with a methanolic solution of Bisphenol A (internal standard). The extract therefrom was analyzed by reverse phase HPLC on a C-18 column using a gradient mobile phase of methanol and methanol: aqueous citric acid buffer and UV detection at 280 nm. The method can be calibrated using a pure sample of Carnosic acid and Carnosol.

The final extract, after centrifuging to remove the fine particulate solids, had 0.49% Carnosol and 5.80% Carnosic acid for a total of 6.2% by weight of active compounds or a two-fold increase in concentration from that of the starting spice. The residual cake solids contained 0.21% Carnosol and 0.23% Carnosic Acid for a total of 0.44% by weight active compounds. This represents about 15% of the active compounds found in the starting spice.

This surprisingly demonstrates that, contrary to prior art, the antioxidant compounds can be efficiently extracted in a continuous, high speed, high pressure process with levels of edible solvent considerably below those disclosed by prior art, providing an extract produced by such a process that is clearly superior in levels of antioxidants. Prior art, in particular U.S. Pat. No. 3,732,111, discloses a method for extraction of Labiatae herbs using prohibitively high ratios of edible solvent in relation to the weight of the starting spice. This necessarily, by simple dilution, results in an extract of low antioxidant concentration that make the extract so weak with respect to the active compounds that it cannot be used in many food systems where the volume of edible solvent in relationship to the compounds of interest is simply too high.

By varying the ratio of edible solvent to ground spice, and the number of countercurrent pressing stages, the levels of active compounds in the extract and residual solids can be controlled in the same manner as described in previous examples.

The starting ground rosemary had an aerobic plate count of 7,000,000. The final extract and residual cake solids had plate counts of 600 and 35,000 respectively. Again, contrary to prior art which suggests that elevated moisture contents and lengthy holding time are necessary, the present process dramatically and rapidly sterilized both the extract and residual solids at such low moisture contents due to the high temperatures, pressures, and shear forces employed.

Example 8: Turmeric Extraction, Standardization, and Sterilization : Using the general procedures as described in the previous Examples, turmeric at 3% moisture and containing 6% curcumin (ASTA method 18.0) was extracted using 10% by weight of Polysorbate 80.RTM., a polyoxyethylene sorbitan ester, in a countercurrent, two-stage pressing operation with the temperature of the residual cake solids being maintained at about 220° F. Curcumin is the pigment responsible for the intense yellow colour of turmeric. The final filtered extract and the residual cake solids are 20% and 3.5% curcumin respectively.

The plate count of the starting spice is about 50,000,000, the final extract 25,000, and the final residual cake solids 1,500,000. Again, contrary to the teachings of prior art, at such low moisture contents.

By varying the amount of edible solvent added prior to the second pressing stage, the levels of curcumin can be accurately and predictably controlled in both the extract and residual cake solids as described in previous examples.

Example 9: Oregano Extraction, Standardization, Activation, and Sterilization : In the same manner as given in previous Examples, ground oregano at 5% moisture is processed in a two-stage, countercurrent extraction process using 10% by weight of soy bean oil. The presses are operated at about 230° F. as measured by the temperature of the cake exiting the pressing stages. The presses are operated at about 20,000 PSI. The plate count of the starting herb is 5,000,000. The plate counts of the final extract and residual cake solids are 800 and 50,000 respectively demonstrating again, contrary to prior art, that the plate counts of spice and herb solids can be significantly reduced at such low moisture contents.

The residual cake solids are compared for flavor strength to the starting ground herb using the method as described in Example 5. Only six of twelve panelists are able to pick the residual cake solids as being less flavourful than the starting ground herb.

Example 10: Other Families and Species of Herb and Spice Plant Materials : In the same general manner as given in previous Examples, other spices and herbs at moisture contents of less than 10% that contain colour, flavor, aroma, or antioxidant principles, are processed and extracted in accord with the method of the present invention.

Such plant materials include, but without limitation, oregano, basil, mint, marjoram, anise, mace, caraway, carrot, garlic, cinnamon, allspice, clove, ginger, cilantro, parsley seed, sesame, cumin, lovage, tarragon, cardamom, fenugreek, mustard, and vanilla, the primary characteristic of such plant materials being that they impart flavor, aroma, colour, or preservative qualities to the food or beverage system in which they are used, and by way of illustration but not by way of limitation, in each case to produce a concentrated extract from the first pressing stage and residual cake solids from the last pressing stage, both standardized with respect to the principal components of interest, both with significantly reduced microbial load, and the residual cake solids having "activated" or quick-releasing principal components of interest in the food or beverage system in which they are used, and for those containing carotenoid pigment, increased stability of the pigment in both the extract and residual cake solids. All this being accomplished under conditions contrary to prior art: without the use of non-edible solvents, low temperatures, and high moisture contents which develop off flavours, aromas, and colours; and without excessive volumes of edible solvents resulting in extracts and residual solids of limited application due to their low content of principal components of interest.

It is thereby seen that an improved countercurrent process for the extraction of spice and herb plant solids using an edible solvent, whereby improved yields of both extract and residual solids are obtained, whereby both the extract and the residual solids have improved stability and freedom from bacterial contamination due to the higher temperatures employed, whereby due to optional advantageous rehydration of residual solids and level of water activity employed an improved stability in the residual solids is attained, whereby an extract in the form of a clear solution can be obtained by removal of gums and particulate solids in the form of their insoluble hydrates, whereby even greater stability can be effected by the employment of edible antioxidants in the solvent utilized, whereby controlled browning of the residual solids may be conveniently effected, and whereby the residual spice or herb plant solids have their tissue ruptured so as to produce quick release of the flavor, aroma, colour, or antioxidant component therein thereby to provide greatly improved use effectiveness, and whereby all products of the process can be conveniently standardized with respect to the principal flavor, aroma, colour, or antioxidant component of interest, all without the expected disadvantages from employing higher temperatures and lower water content as clearly indicated by the prior art, and whereby all of the stated objects of the invention have been accomplished, has been provided.

It is to be understood that the invention is not to be limited to the exact details of operation, or to the exact compositions, methods, procedures, or embodiments shown and described, as obvious modifications and equivalents will be apparent to one skilled in the art, and the invention is therefore to be limited only by the full scope which can be legally accorded to the appended claims.

Aromatherapy

Basic Principles of Aromatherapy

Essential oils are obtained by 2 main methods:

(1). Expression (also called pressing) *i.e.* cold pressed lemon oil.

(2). Distillation, either steam, water or dry.

For oils such as Camphor it is processed three times to produce the three types of oil. The first produces Brown Camphor, the second Yellow Camphor and the third White Camphor.

The other methods for extraction are:

(1). Solvent. This produces a 'Concrete', a 'Resinoid' and an 'Absolute'.

(2). Enfleurage or Pomade. This method for producing Essential oils is not used much any more, as it is an expensive and time consuming process.

A Concrete is obtained through the use of a hydrocarbon solvent to extract the Essential oil from the plant matter. This is used for the Essential oils such as Rose, Jasmine and Ylang Ylang. The Ylang Ylang concrete is approx. 80% Essential oil and 20% wax. Jasmine is approx. 50% Essential oil and 50% wax. A second extraction may also be performed to the plant matter using alcohol which produces an Absolute such as Neroli.

A Resinoid is obtained by the same method, but it is produced from resin based plants such as Amber and Frankincense. A Pomade was obtained by the use of layers of fat onto which the petals of plants such as Tuberose and Jasmine were laid out and left to dry. The fat collected the Essential oils which were later extracted. This process has now been replaced by solvent extraction.

Essential oils are found in very small quantities in many plants. For example: for every 100 kg of plant matter the following plants produce these amounts of oil:

Eucalyptus	-	3kg
Lavender	-	1.9kg
Ylang Ylang	-	1.6-2kg
Juniper	-	1/2-1.2kg
Rose	-	0.05kg

As you can see it takes a lot of plants to make a small amount of Essential oil.

Plants from different areas/countries can be more expensive (lower % of oil per plant) or cheaper (higher % of oil per plant).

Essential oils are categorized by:

(a) Their plant type *i.e.* Citrus

(b) By their Note. *i.e.* Base note

The plant types are: Citrus, floral, herb, spice, wood and resin. Although the oils can blend with any other oils, they blend better with oils of their own group, or with an oil of a similar group. *i.e.* Lemon (citrus) & Lavender (floral) and Cedarwood (wood) & Patchouli (herb).

The note types are: Top, Middle and Base. Top note oils such as Neroli and Lemon evaporate and lose their aroma quickly when left open. A top note oil will last approx. 1 week if left opened. A middle note oil such as Lavender and Geranium is slightly more stable and will evaporate and lose its aroma approx. 2-3 weeks when left opened. Base note oils are the heavy oils such as Sandalwood and Patchouli which will evaporate much more slowly, taking about a month. When you smell a blended oil you can usually pick which oils are the top, middle and base oils as the first one you notice will be a top note. The next scent you notice will be from the middle notes and the heavy, lingering scent will belong to the base note. In this way you can tell which oil is which note.

A good blend will contain at least one of each note, to add a layered effect to the blend.

Synthetic vs. Natural Oils

Most of you will have seen cheap bottles of 'fragrant' oils or bottles of 'blended Essential oils' as well as bottles of 'Pure Essential oils'. Each one has it's own uses.

Fragrant Oils: These are the synthetic oils.

Scientists have been able to reproduce about 90% of the natural occurring Essential oil. This is what a fragrant oil is. It is the remaining 10% of the natural Essential oil that holds the therapeutic content of the oil, and that 10% is what distinguishes the two. Fragrant oils have NO therapeutic content, they smell nice (sometimes better than the Essential oils!) but that is all that they should be used for. They work well in oil burners, in baths, in pot pourri, as a perfume and various other uses, but will have no healing effect other than to smell nice.

Blended Essential Oils: These are made from Pure Essential oils, but have been diluted with a 'Carrier oil', usually Sweet Almond, jojoba, Safflower or Apricot Kernel.

These oils are blended because Essential oils are too strong to apply directly to the skin, they must be diluted first. The Blended oil has been diluted so it is safe to apply directly onto the skin. These have therapeutic qualities, which are usually printed on the label. They may also be blended to make them cheaper, such as Rose, Neroli, Jasmine, Chamomile and Ylang Ylang.

Pure Essential Oils: These are the Essential oils in their 100% pure state.

These oils should never be applied straight onto the skin, and many Essential oils will have safety data printed onto the label if they have any harmful effects. (there are a few exceptions to the "never put oils directly on the skin" rule... which I will go into later) When purchasing an Essential oil there are 4 ways to help tell if you are buying the pure oil:

(1) Look for the words "100%" and "Pure Essential oil". This is only a guide as many brands of fragrant oil have 100% pure written on them, and some brands of Essential oils (such as the brand I use) don't have 100% pure on the label. There are also sneaky people who use the "100% Pure Essential Oil" as their brand name....(and if the word 'fragrant' appears anywhere, chances are it isn't an Essential oil.)

(2) Smell the oils, Pure Essential oils smell like the plant it comes from. If the scent is slightly fake, very sweet or in the case of Rose really strong, it is more likely to be a fragrant oil.

(3) Look at the price, Pure Essential oils are expensive, due to the quantity of plants needed to produce the oil (for *e.g.* it takes approx. 15 roses to make 1 drop of Rose oil.) Each oil should have a different price, as some are cheaper to produce than others. As a general rule... if they are under $5 (Australian money... not sure of the prices in other countries) they are probably either blended or a fragrant oil.

(4) There are no Essential oils of Strawberry, Dewberry, Rainforest or Nanna's Garden. If something like these is packaged the same and on the same stand, then chances are all the oils are synthetic oils.

The Common Essential Oils

The following information has been gathered from many books over many years. Some oils work by promoting balance, which is why they are listed as working well with things such as oily skin and dry skin. What this means is that if your skin is oily, it will help dry it out, if it is dry it will help add moisture. In other words, it will balance.

Different sources will give different properties to the Essential oils - some may even contradict others. It is for this reason that it is best to leave Aromatherapy to trained professionals. The following information is not intended to prescribe or diagnose in any way. It is not meant to be a substitute for professional medical advice/ assistance. The intention of this list is to give information on the historical usage of Aromatherapy. Please consult your doctor or other health professional if you have any medical complaints. Do not attempt to self diagnose or self prescribe herbal treatments for yourself or others. The Author (Obsidian) shall not be responsible for any misuse or abuse of this information.

Aniseed

Botanical Name: Pimpinella Anisum

Note - Top to middle

Type - Spice

Family - Umbelliferae

Part - Herbs/seeds

Extraction - Steam Distillation

Aroma: Sweet and spicy, pungent, Licorice like, very warming.

Blends well With: Amyris, Bay, Cardamon, Caraway, Cedarwood, Coriander, Dill, Fennel, Galbanum, Mandarin, Petitgrain, Rose wood.

Contraindications: Sensitive skin, Pregnancy, Over use can cause sluggishness, drowsiness and dizziness. In extreme cases can cause cerebral congestion and circulatory problems.

Properties: Carminative, Antiseptic, Antispasmodic, Decongestant, Parasiticide, Antiemetic, Aphrodisiac, Cardiac, Digestive aid, Diuretic, Expectorant, Galactagogue, Insecticide, Laxative, Parturient, Pectoral stimulant, Stomachic.

Uses: Respiratory problems, Digestive problems, Flatulence, Indigestion, Period pains, Stimulate lactation, Lice, Scabies, Muscular aches and pains, Rheumatism, Bronchitis, Coughs, Colic, Cramp, Colds, Stimulate the mind, Dyspepsia, Nausea, Vomiting, Stimulating peristalsis, Oliguria (low quantity of urine), Stimulates cardiac fatigue, Lice, Scabies, Infectious skin diseases.

Cautions: Not generally recommended for use in the home as prolonged or over use may be harmful.

Basil (Sweet)

Botanical Name: Omicum Basilicum

Note - top

Type - Herb

Family - Labiatae

Part - Flowering top

Extraction - Steam Distillation

Aroma: Very clear, sweet and slightly spicy.

Blends well With: Bergamot, Black pepper, Citronella, Chamomile, Clary Sage, Geranium, Hyssop, Lavender, Lemongrass, Lime, Marjoram, Melissa, Neroli, Peppermint, Rose, Sandalwood, Verbena.

Contraindications: Pregnancy, Sensitive skin, Do not use to excess.

Properties: Soothing, Uplifting, Antiseptic, Insect repellent, Digestive aid, Antispasmodic, Emmenagogue, Analgesic, Antidepressant, Antispasmodic, Antivenomous, Aphrodisiac, Bactericide, Carminative, Cephallic, Expectorant, Febrifuge, Galactagogue, Insecticide, Nervine, Stomachic, Sudorific, Tonic, Restorative, Stimulant, Vermifuge, Effects adrenal cortex, Imitates oestrogen, Minimise uric acid in muscles, stimulates blood flow, Helps expel afterbirth, Cleanse intestines and kidneys.

Uses: Fatigue, Depression, Respiratory infections, Colds, Cough, Bronchitis, Asthma, Sinusitis, Fever, Gout, Indigestion, Insect bites and stings, Breast engorgement, Fainting, Flatulence, Insomnia, Muscular aches and pains, Rheumatism, Anxiety, Migraine, Nervous tension, Sharpening senses, Nerve tonic, Concentration, Hysteria, Headaches, Temporary paralysis, Nasal polyps, Earache, Allergies, Bronchitis, Emphysema, Flu, Whooping cough, Catarrh, Vomiting, Gastric spasm, Nausea, Dyspepsia, Hiccups, Menstrual problems, Scanty periods, Fertility problems, Fatigue, Stress, Poor circulation.

Bay

Botanical Name: Laurus Nobilis

Note - top

Type - Spice

Family - Lauraceae

Part - Leaves and twigs

Extraction - Steam Distillation

Aroma: Sweet and spicy, similar to Cinnamon

Blends well With: Citrus Oils, Cedarwood, Clary Sage, Coriander, Cypress, Eucalyptus, Ginger, Hyssop, Juniper, Lavender, Lemon, Marjoram, Myrtle, Orange, Rose, Rosemary, Thyme, Ylang Ylang.

Contraindications: Sensitive skin, Pregnancy, Irritate mucous membranes.

Properties: Analgesic, Antineuralgic, Antiseptic, antispasmodic, Aperitif, Astringent, Cholagogue, Diuretic, Emmenagogue, Febrifuge, Hepatic, Insecticides, Parturient, Stimulant, Stomachic, Sudorific, Tonic, Mildly narcotic, Warms emotions, Reproductive tonic, Uplifting.

Uses: Colds, Flu, Bronchitis, Digestive aid, Dyspepsia, Flatulence, Appetite stimulant, Settles stomach, Liver and Kidney tonic, Rheumatism, Aches and pains, Sprains, Fever, Infectious disease, Scanty periods, Speeds up childbirth, Ear infections, Dizziness, Restores balance, Minor respiratory problems, Indigestion.

Cautions: People prone to sensitive or allergy prone skin should not use Bay oil. Use in moderation and never undiluted.

Bergamot

Botanical Name: Citrus Bergamia

Note - top

Type - Citrus

Family - Rutacaea

Part - Peel

Extraction - Expression

Aroma: Light, delicate, refreshing, something like Orange and lemon with slight floral undertones.

Blends well With: Basil, Cardamon, Chamomile, Coriander, Cypress, Eucalyptus, Geranium, Juniper, Jasmine, Lavender, Lemon, Marjoram, Mimosa, Myrtle, Neroli, Palmarosa, Patchouli, Petitgrain, Sandalwood, Ylang ylang.

Contraindications: Phototoxic, sensitive skin.

Properties: Antiseptic, Parasiticide, Antidepressant, Analgesic, Antispasmodic, Carminative, Cicatrisant, Cordial, Deodorant, Digestive, Expectorant, Febrifuge, Insecticide, Sedative, Stomachic, Tonic, Vermifuge, Vulnerary, Keeps pets away from plants, Uplifting, Vitalising, Anti-inflammatory,

Uses: Dandruff, Urinary and Respiratory infections, Skin infections, Throat and mouth infections, Scalp and skin, Psoriasis, Acne, Ulcers, Stress related conditions, Depression, Coldsores, Shingles, Insomnia, Anxiety, Stress, Bad breath, Deodorant, Worms, Intestinal parasites, Colic, Appetite stimulant, Vaginal infections, Flatulence, Boils, Eczema, Varicose Ulcers, Wounds, Tonsillitis, Colds, Fever, Flu, Abscesses, Cystitis, Digestive aid, Eczema, Insect repellent,

Oily skin, Infectious diseases, PMS, Dyspepsia, Flatulence, Colic, Indigestion, Gallstones, Anorexia, Breathing difficulties, TB, Chickenpox, Uterine Tonic, STDs, Gastro-intestinal spasm, Cellulite, Sore throat, Bronchitis.

Black Pepper

Note - Middle

Type - Spice

Family - Piperaceae

Part - fruit

Extraction -Distillation.

Aroma: Sharp and Spicy.

Blends well With: Basil, Bergamot, Cedarwood, Cypress, Frankincense, Geranium, Grapefruit, Juniper, Lemon, Marjoram, Palmarosa, Rosemary, Sandalwood, Ylang Ylang.

Contraindications: Kidney Problems, Sensitive Skin.

Properties: Analgesic, Antiemetic, Antiseptic, Antispasmodic, Aphrodisiac, Cardiac, Carminative, Detoxicant, Digestive, Diuretic, Febrifuge, Laxative, Rubefacient, Stimulant, Stomachic, Tonic, Helps frustration, Warms the heart with there is indifference, Dilates local blood vessels, Stimulates peristalsis, Stimulates circulation, Helps form new blood cells, Antidote for fish and mushroom poisoning, Reduces fat, Increases flow of Saliva, Stimulates appetite, Expels wind, Fortifying stomach, Expels toxins.

Uses: Muscle aches and pains, Tired and aching limbs, Muscle stiffness, Rheumatoid Arthritis, Temporary Paralysis, Vomiting, Bowel problems, Anaemia, Respiratory illnesses, Fever (Small amounts).

Cajuput

Note - top

Type - Herb

Family - Myrtacaea

Part - leaves and twigs

Extraction - Distillation

Aroma: Sweet, herbaceous and rather penetrating

Blends well With: Angelica, Bergamot, Birch, Cardamon, Cedarwood, Clove, Eucalyptus, Geranium, Immortelle, Lavender, Myrtle, Niaouli, Nutmeg, Pine, Rose, Rose wood, Thyme

Contraindications: Use with caution, sensitive skin, irritate mucous membranes

Properties:Analgesic, Antidontalgic, Antineuralgic, Antirheumatic, Antiseptic, Antispasmodic, Balsamic, Cicatrisant, Decongestant, Expectorant, Febrifuge, Insecticide, Pectoral, Stimulant, Sudorific, Vermifuge, Clears the mind, Balances mind and body, Imitates Oestrogen.

Uses: Respiratory tract infections, Fever, Colds, Pharyngitis, Laryngitis, Bronchitis, Asthma, Ease Chronic Pulmonary Disease, Colic, Inflammation of the intestines, Enteritis, Dysentery, Gastric spasm, Nervous vomiting, Intestinal Parasites, Urinary infections, Cystitis, Urethritis, Neuralgia, Headaches, Toothache, Earache, Gout, Chronic rheumatism, Muscle Stiffness, Muscle aches and pains, Menopausal problems.

Cedarwood

Note - Base

Type - Wood

Family - Cupressacae/Pinaceae

Part - Wood

Extraction - Distillation

Aroma: Woody, reminiscent of sandalwood but slightly "drier" and almost pine tones.

Blends well With: Benzoin, Bergamot, Black pepper, Cajuput, Cinnamon, Cypress, Frankincense, Ginger, Jasmine, Juniper, Lavender, Lemon, Linden, Myrrh, Neroli, Patchouli, Pine, Rose, Rosemary, Sandalwood, Vetiver, Ylang Ylang.

Contraindications: Sensitive Skin, Pregnancy.

Properties: Calming, Soothing, Antiseptic, Astringent, Diuretic, Emollient, Expectorant, Fungicide, Insecticide, Sedative, Tonic, Better for chronic (long standing) problems than acute (recent) ones, Balances body and mind, Insect repellent, Aphrodisiac.

Uses: Nervous Tension, Anxiety, Meditation, Bronchitis, Coughs, Catarrh, Excess phlegm, Cystitis, Genito-Urinary tract problems, Kidney problems, Rheumatism, Arthritis, Oily skin, Acne, Clearing scabs, Dermatitis, Psoriasis, Hair tonic, Seborrhoea, Dandruff, Alopecia, Softening skin (especially when mixed with Cypress and Frankincense), Respiratory infections, Glandular problems, Stress, Insect bites and stings, Dermatitis, Eczema, Itching, Coughs, Fungal infections, Hair loss, Ulcers, Bronchitis, Catarrh, PMT, Tiredness, Vaginal infections, Urinary tract infections, Congestion, Sinusitis.

Chamomile

Note -Middle

Type - Flower

Family - Compositae

Part - Flowers

Extraction - Distillation

Aroma: Fruity, apple-like.

Blends well With: Angelica, Basil, Benzoin, Bergamot, Clary sage, Geranium, Jasmine, Lavender, Lemon, Marjoram, Neroli, Palmarosa, Patchouli, Rose, Star Anise, Ylang Ylang.

Contraindications: Pregnancy (early Months)

Properties: Emmenagogue, Analgesic, Antiallergenic, Anticonvulsive, Antidepressant, Antibacterial, Antiemetic, Antiphlogistic, Antipruritic, Antirheumatic, Antiseptic, antispasmodic, Carminative, Cholagogue, Cicatrisant, Digestive, Diuretic, Emollient, Febrifuge, Hepatic, Nervine, Sedative, Splenetic, Stomachic, Sudorific, Tonic, Vermifuge, Vulnerary, Soothing, Calms the mind, Regulates menstruation, Stimulates growth of white corpuscles.

Uses: Anxiety, Tension, Anger, Fear, Promotes relaxation, Gives patience and peace, Allay worry, Insomnia, Muscular pain, Lower back pain, Headache, Neuralgia, Toothache, Earache, Menstrual problems, Period pain, PMT, Menopause, Soothes stomach, Gastritis, Diarrhoea, Colitis, Peptic Ulcers, Vomiting, Flatulence, Irritated bowels, Liver problems, Jaundice, Genito-urinary problems, Anaemia, Burns, Blisters, Inflamed wounds, Ulcers, Boils, Dermatitis, Acne, Herpes, Psoriasis, Broken Capillaries, Improves skin elasticity, Itching, Puffiness, Strengthen tissues, Skin cleanser, Hair tonic, Cramps, Nappy Rash, Nervous Tension, Neuralgia, Bleaching hair, Cracked nipples, Children's tantrums, Fevers, Insomnia, Mastitis.

Cinnamon

Note - Base

Type - Spice

Family - Lauracaea

Part - Bud/Bark/leaf

Extraction - Distillation

Aroma: Spicy, Sharp, Sweet and Musky

Blends well With:Benzoin, Cardamon, Clove, Coriander, Eucalyptus, Frankincense, Galbanum, Ginger, Grapefruit, Lavender, Lemon, Mandarin, Orange, Pie, Rosemary, Thyme.

Contraindications: (Leaf is safer as Bud and Bark can irritate skin more) Use with care in small doses, Pregnancy, High doses could cause convulsions

Properties: Anaesthetic, Antidontalgic, Antiseptic, Antiputrefative, Antispasmodic, Aphridesiac, Astringent, Cardiac, Carminative, Emmenagogue, Escharotic, Haemostatic, Insecticide, parasiticide, Sialogue, Stimulant, Stomachic, Vermifuge, Stimulates tears, Stimulates saliva and mucous, Stimulates secretion of gastric juices, Stimulates Circulatory system.

Uses: Respiratory infections, Colds, Flu, Breathing difficulty, Fainting, Infectious diseases, Viral infections, Intestinal infections, Digestive spasm, Asthenia, Dyspepsia, Colitis, Flatulence, Gastralgia, Diarrhoea, Nausea, Vomiting, Cholera, Typhoid, Period pain, Scanty menstruation, Leucorrhoea, Impotence, Muscle spasm, Rheumatism, Insect bites, Respiratory and digestive problems, Lice, Scabies, Tooth and Gum care, Warts, Wasp stings, Poor circulation, Rheumatism, Anorexia, Colitis, Aids childbirth, Frigidity, Cough, Depression, Flu.

Citronella

Note - Top

Type - Citrus

Family - Graminae

Part - Grass

Extraction - Distillation

Aroma: Sweet and Lemony

Blends well With: Bergamot, Cajuput, Cedarwood, Eucalyptus, Geranium, Lavender, Lemon, Mandarin, Neroli, Orange, Peppermint, Petitgrain, Pine, Sage, Ylang Ylang.

Contraindications: May irritate Sensitive Skin, Pregnancy.

Properties: Antidepressant, Antiseptic, Deodorant, Insecticide, Parasiticide, Tonic for heart and nervous system, Stimulant, Clearing mind, uplifting.

Uses: Insect repellent, Excessive perspiration, Oily skin, Cold, flu, Minor infections, Fatigue, Headache, Migraine, Neuralgia, Tired feet, Rheumatic aches and pains, Softens skin.

Clary Sage

Note - Top to Middle

Type - Herb

Family - Labiatae

Part - Herb/Flowering tops and leaves

Extraction - Distillation

Aroma: Heavy herby and nutty.

Blends well With: Angelica, Basil, Bay, Bergamot, Cardamon, Cedarwood, Citronella, Coriander, Cypress, Frankincense, Geranium, Grapefruit, Jasmine, Juniper, Lavender, Lemon, Lime, Myrtle, Petitgrain, Rose, Sandalwood, Ylang Ylang.

Contraindications: Not before driving or operating heavy machinery, Not with Alcohol consumption, Large doses can cause headaches, Pregnancy, Epilepsy

Properties: Anticonvulsive, Antidepressant, Antiphlogistic, Antiseptic, Antispasmodic, Antisudorific, Aphrodisiac, Balsamic, Carminative, Deodorant, Digestive, Emmenagogue, Hypotensive, Nervine, Parturient, Sedative, Stomachic, Tonic, Uterine, hormone balancing, Kidney tonic, Soothing, Uplifting, Balance, Encourages hair growth, Astringent, Anti-inflammatory, Antibacterial, Powerful muscle relaxant.

Uses: Nervous tension, Anxiety, Uterine problems, Scanty periods, PMT, Menstrual cramps, Stress, Fertility, Digestive problems, Flatulence, Gastric spams, Headaches, Migraine, Tension, cramp, Excessive perspiration, TB, Asthma, Sore throats, Throat infections, Brings vigour after illness, Convalescence, Overcoming drug addictions, Depression and hopelessness, Cell regenerate, Scalp problems, Dandruff, Oily hair, Inflamed and puffy skin, Hypertension, Colds, Menstrual problems, Dry skin, Digestive problems.

Clove

Note -Base

Type - Spice

Family - Myrtaceae

Part - (Tree) Bud or Leaf

Extraction - Distillation

Aroma: Spicy, strong.

Blends well With: Basil, Benzoin, Black pepper, Cinnamon, Citronella, Grapefruit, Lemon, Nutmeg, Orange, Peppermint, Rosemary.

Contraindications: (Highly irritant), Sensitive skin, Pregnancy, Not for use with children

Properties: Analgesic, Anaesthetic, Antidontalgic, Antiemetic, Antineuralgic, Antiseptic, Antispasmodic, Aperitif, Aphrodisiac, Carminative, Caustic, Cicatrisant, Disinfectant, Insecticide, Parturient, Splenetic, Stimulant, Stomachic, Uterine, Vermifuge, Expectorant, Antihistamine, Aids digestion, Antidepressant, Parasiticide, Tonic for the Kidneys; spleen and stomach, Antibacterial.

Uses: Dyspepsia, Gum infections, Toothache, Bronchitis, Catarrh, Scabies, Athletes foot, Cold, Flu, Tinea, Acne, Bruises, Burns, Cuts, Ulcers, Wounds, Arthritis, Rheumatism, Sprains, Asthma, Colic, Nausea, Minor infection, Diarrhoea, Flatulence, General weakness, Muscular aches and pains, Tension, Scar tissue, Memory, Lethargy, Vomiting, Intestinal spasm, Halitosis, Headache, Respiratory problems, TB, Impotence, Frigidity, Helps child birth pains, Lupus.

Cypress

Note - Middle to base

Type - Wood

Family - Cupressaceae

Part - (Tree) Leaves and cones

Extraction - Distillation

Aroma: Woody and spicy, but refreshing and clear

Blends well With: Benzoin, Bergamot, Clary Sage, Juniper, Lavender, Lemon, Linden, Orange, Pine, Rosemary, Sandalwood.

Contraindications: High Blood pressure, Pregnancy,

Properties: Antiseptic, Astringent, Aids healing, Insecticide, Vasoconstricting, Tonic for circulatory system and liver, Cicatrisant.

Uses: Urinary problems, Fluid retention, Excessive perspiration, Diarrhoea, Menorrhagia, Tired aching legs, Oily skin, Rheumatism, Back ache, Haemorrhage, Swelling, Nasal congestion, Cold, Flu, Scars, Wounds, Haemorrhoids, Varicose veins, Asthma, Cellulitis, Menstrual cramp, Poor circulation, Spasmodic coughs, Dysmenorrhoea, Stress, Nervous tension, Acne, Eczema, Gum disorders, Incontinence, Menopausal problems, Nose bleeds, Whooping cough.

Eucalyptus

Note - Top

Type - Wood

Family - Eucalyptus globulus

Part - (Tree) leaves

Extraction - Distillation

Aroma: Clear, sharp and piercing

Blends well With: Benzoin, Cajeput, Cedarwood, Coriander, Cypress, Juniper, Lavender, Lemon, Lemongrass, Marjoram, Melissa, Peppermint, Pine, Rosemary, Star Anise, Teatree, Thyme.

Contraindications: Sensitive skin, Kidney problems, Toxic if taken internally, Heart problems, Not for children, High Blood pressure, Pregnancy, Epilepsy, Do not use if you are on any Homoeopathic medication.

Properties: Analgesic, Antirheumatic, Antipholgistic, Antiseptic, Antispasmodic, Antiviral, Antibacterial, Balsamic, Cicatrisant, Decongestant, Deodorant, Depurative, Diuretic, Expectorant, Febrifuge, Hypoglycemiant, Insecticide, Rubefacient, Stimulant, Vermifuge, Vulnerary, Parasiticide, Cools and soothes, Strengthens nervous system,

Uses: Respiratory complaints, Croup, Bronchitis, Ringworm, Insect bites, Shingles, Chicken pox, Herpes, Muscle aches and pains, Dysentery, Hay fever, Burns, Throat infections, Colds, Chest infections, Acne, Asthma, Boils, Cold sores, Cuts, Fever, Flu, Lice, Laryngitis, Rheumatism, Skin infections, Sore throat, Urinary infections, Aids concentration, Catarrh, TB, Sinusitis, Migraine, Typhoid, Diphtheria, Malaria, Cystitis, Diarrhoea, Gall stones, Nephritis, Gonorrhoea, Diabetes, Haemorrhage, Neuralgia, Pyorrhoea.

Fennel

Note - Top to middle

Type - Herb

Family - Umbelliferae

Part - (Herb) seed

Extraction - Distillation

Aroma: Floral, herbal and spicy

Blends well With: Basil, Geranium, Lavender, Lemon, Rose, Rosemary, Sandalwood.

Contraindications : Pregnancy, Epilepsy, Toxic if used in large doses, not for babies.

Properties : Antiphlogistic, Antiseptic, Antispasmodic, Aperitif, Carminative, Detoxicant, Diuretic, Emmenagogue, Expectorant, Galactagogue, Insecticide, Laxative, Resolvent, Splenetic, Stimulant, Stomachic, Sudorific, Tonic, Vermifuge, Reduces lactobacillus, Has an action similar to Oestrogen, Stimulates lactation, Reduces the toxic effect of alcohol, Body cleansing and detoxifying.

Uses : Kidney Stones, Menstrual problems, PMT, Respiratory problems, Urinary tract infections, Flatulence, colic, Menopausal problems, Diuretic, Gout, Liver problems, Children's complaints, Intestinal parasites, Nausea, Vomiting, Indigestion, Cellulitis, Obesity, Odema, Rheumatism, Asthma, Bronchitis, Anorexia, Bruises, Dull skin, Oily skin, Alcohol poisoning, Appetite reduction, Arthritis, Colitis, Constipation, Fluid retention, Mouth and gum problems, Hangovers, Stomach ailments, Eases digestion, Hiccups, Whooping cough, Scanty periods.

Frankincense

Note - Middle to base

Type - Resin

Family - Burseraceae

Part - (Tree) Bark

Extraction - Distillation

Aroma: Woody, spicy, and a hint of lemon

Blends well With: Basil, Black Pepper, Cedarwood, Cinnamon, Citrus oils, Galbanum, Geranium, Ginger, Grapefruit, Lavender, Orange, Melissa, Myrrh, Neroli, Patchouli, Pine, Sandalwood, Vetiver.

Contraindications: Pregnancy.

Properties: Uterine tonic, Astringent, Blood coagulant, Calming, Deepens breathing, Induces sweating, Pulmonary antiseptic, Calming, Clears the lungs, Anti-inflammatory.

Uses: Bronchitis, Laryngitis, Asthma, anxiety, Stress, Emotional upsets, Urinary tract infections, Cystitis, Uterine tonic, Wounds, Mature skin, Blemishes, Dry skin, Scars, Catarrh, Dysmenorrhoea, Cold, Flu, Nervous tension, Acne, Coughs, Laryngitis, Meditation, Menstrual problems, PMS, Menorrhagia, Respiratory conditions, Scar tissue, Improves skin tone, genito-urinary infections, Catarrh,

Nephritis, Genital infections, Uterine haemorrhage, Post natal depression, Calms during labour, Breast inflammation, Dyspepsia, Belching, Wrinkles, Oily skin, Ulcers, Carbuncles.

Geranium

Note - Middle

Type - Floral

Family - Geraniacea

Part - (plant) Flowers and leaves

Extraction - Distillation

Aroma: Sweet and heavy, a little like rose with a hint of mint.

Blends well With: Angelica, Basil, Bay, Bergamot, Carrot Seed, Cedarwood, Citronella, Clary Sage, Grapefruit, Hyssop, Jasmine, Lavender, Lemon, Lime, Mandarin, Marjoram, Neroli, Orange, Patchouli, Petitgrain, Rose, Rosemary, Sandalwood, Tea Tree.

Contraindications: Sensitive skin, Pregnancy.

Properties: Analgesic, Anticoagulant, Antidepressant, Aphrodisiac, Antiseptic, Anti-fungal, Anti-inflammatory, Astringent, Cicatrisant, Cytophylactic, Diuretic, Deodorant, Haemostatic, Hypoglycemiant, Insecticide, Styptic, Tonic, Vasoconstrictor, Vulnerary, Cleanses the body of toxins, Stimulates lymphatic system, Clears digestive mucous, Balances Sebum, Skin cleanser, Improves flow of blood.

Uses: Anxiety, Depression, Balances mind, Stress, PMT, Menorrhagia, Lack of vaginal secretions, Inflammation, Breast Congestion, Congested system, Jaundice, Kidney stones, Gallstones, Diabetes, Urinary infections, Fluid retention, Swollen ankles, Throat infections, Mouth Infections, Neuralgia, Gastritis, Colitis, Insect Repellent, Eczema, Shingles, Herpes, Ringworm, Chilblains, Oily skin, Sluggish skin, Congested skin, Cuts, Menopausal conditions, Cellulitis, Mastitis, Emotional problems, tonsillitis, Burns, Muscular aches and pains, Respiratory conditions, Rheumatism, Swelling, Ulcers, Apathy, Acne, Bruises, Broken capillaries, Dermatitis, Haemorrhoids, Lice, Mature skin, Mosquito repellent, Wounds, Engorgement of breasts, Poor circulation, Sore Throat, Nervous tension, Abscesses, Boils, Bronchitis, Indigestion, Insect bites.

Jasmine

Note - Middle to Base

Type - Floral

Family - Jasminacaea

Part - Flowers

Extraction - Enfleurage/Solvent Extraction

Aroma: Sweet and flowery.

Blends well With: Bergamot, Citrus oils, Chamomile, Clary Sage, Coriander, Frankincense, Geranium, Guaiacwood, Immortelle, Lavender, Mandarin, Melissa, Mimosa, Myrtle, Neroli, Orange, Palmarosa, Patchouli, Peppermint, Petitgrain, Rose, Rose wood, Sandalwood, Vetiver, Ylang Ylang.

Contraindications: Sensitive skin, Pregnancy, Not for Babies, Excessive use can cause problems with bodily fluids such as phlegm, May have a narcotic effect and cause lack of concentration.

Properties: Menstrual regulator, Sedative, Antidepressant, Stimulates lactation, Boosts confidence, Analgesic, Balances hormones, Increases male fertility.

Uses: Stress, Skin Care, Throat infections, Coughs, Catarrh, Listlessness, Apathy, Melancholy, Hopelessness, Menstrual cramps, Back pain, Labour Pain, Impotence, Frigidity, Dry Skin, Itchy skin, Muscle aches and pains, Child Birth, Lethargy, Nervous tension, PMS, Post natal depression, Vaginal infections, Premature ejaculation, Hoarseness.

Juniper

Note - Middle

Type - Herb

Family - Cupressacea

Part - Berries

Extraction - Distillation

Aroma: Slightly woody

Blends well With: Benzoin, Bergamot, Cedarwood, Citrus oils, Cypress, Frankincense, Geranium, Ginger, Grapefruit, Orange, Lavender, Lemongrass, Lime, Melissa, Pine, Rosemary, Sandalwood, Vetiver.

Contraindications: Pregnancy, Kidney problems, not for use with babies.

Properties: Antiseptic, Anti-rheumatic, Antispasmodic, Aphrodisiac, Astringent, Carminative, Cicatrisant, Clears Mucous from intestines, Depurative, Detoxicant, Digestive aid, Disinfectant, Diuretic, Emmenagogue, Nervine, Insecticide, Parturient, Regulating appetite, Regulating periods, Rubefacient, Stimulating, Stomachic, Sudorific, Tonic, Vulnerary.

Uses: Genito-Urinary infections, Cystitis, Strangury (Inability to pass urine), Kidney stones, Cellulitis, Dropsy, Fluid retention, releasing toxins, Piles, Obesity, Cirrhosis, Arthritis, Rheumatism, Gout, Sciatica, Stiff joints, Menstrual cramps, Childbirth, Problem skin, Oily skin, Congested skin, Seborrhoea, Acne, Blocked pores, Dermatitis, Eczema, Psoriasis, Swelling, Haemorrhoids, Ulcers, Stress, Oily Hair, Indigestion, Insomnia, Menstrual problems, Loss of appetite, Muscular aches and pains, PMS, Intestinal parasites, Colds, Flu, Anxiety, Purifying the blood, Circulation problems, Cough, Diarrhoea, Fatigue,

Lavender

Note - Middle

Type - Flower

Family - Labiatae

Part - Flowers

Extraction - Distillation

Aroma: Floral and slightly woody

Blends well With: Basil, Bay, Bergamot, Cedarwood, Chamomile, Citronella, Clary Sage, Cypress, Eucalyptus, Geranium, Grapefruit, Hyssop, Jasmine, Lemon, Lemongrass, Lime, Mandarin, Marjoram, Myrtle, Neroli, Nutmeg, Orange, Patchouli, Peppermint, Petitgrain, Pine, Thyme, Rosemary, Tea Tree.

Contraindications: Low Blood pressure, Early Pregnancy.

Properties: Analgesic, Anticonvulsive, Antidepressant, Antiphlogistic, Antirheumatic, Antiseptic, Antispasmodic, Antiviral, Bactericide, Balances central nervous system, Carminative, Cholagogue, Cicatrisant, Cordial, Cytophylactic, Decongestant, Deodorant, Detoxicant, Diuretic, Emmenagogue, Fungicide, Hypotensive, Increases gastric secretions, Nervine, Promotes growth of new skin cells, Restorative, Sedative, Splenetic, Stimulates bile production, Sudorific, Vulnerary.

Uses: Anger, Exhaustion, High blood pressure, Heart palpitations, Insomnia, Muscular spasm, Sprains, Strains, Rheumatic pains, Bronchitis, Asthma, Catarrh, Colds, Laryngitis, Throat infections, TB, Infections, Period pain, Scanty periods, Leucorrhoea, Childbirth,

Cleanse spleen and liver, Nausea, Vomiting, Colic, Flatulence, Insect repellent, Burns, Sunburn, Acne, Eczema, Psoriasis, Abscesses, Boils, Carbuncles, Fungal growths, Swelling, Scars, Gangrene, Alopecia, Headaches, Lice, fleas, Appetite stimulating, Whooping cough, Flu, Nervous Tension, Stress, Vertigo, Shock, Athlete's foot, Genito-urinary problems, Cystitis, Scalds, Wounds, Sores, Varicose veins, PMT, Ulcers, Spider bites, Ringworm.

Lemon

Note - Top

Type - Citrus

Family - Rutaceae

Part - Peel

Extraction - Expression/Distillation

Aroma: Citrus, Sharp and fresh

Blends well With: Bay, Benzoin, Black pepper, Cardamon, Chamomile, Citronella, Citrus oils, Eucalyptus, Fennel, Frankincense, Geranium, Ginger, Jasmine, Juniper, Lavender, Linden, Neroli, Peppermint, Rose, Sandalwood, Ylang-Ylang.

Contraindications: Sensitive skin, Phototoxic, High Blood pressure.

Properties: Antacid, Antisclerotic, Antiscorbutic, Antineuralgic, Antirheumatic, Antipruritic, Antiseptic, Astringent, Bactericide, Carminative, Cicatrisant, Decongestant, Depurative, Diuretic, Emollient, Escharotic, Febrifuge, Haemostatic, Hepatic, Hypoglycemiant, Hypotensive, Insecticide, Laxative, Stomachic, Tonic, Vermifuge.

Uses: Varicose veins, Easing blood flow, High Blood pressure, Anaemia, Restoring vitality to Red blood cells, Stimulating white corpuscles, Infectious diseases, Nosebleeds, Sore throat, Coughs, Colds, Flu, Fever, Cold sores, Herpes, Digestive complaints, Counteracting acidity in stomach, Diabetes, Cleanses body, Constipation, Cellulite, Headaches, Migraine, Neuralgia, Rheumatism, Arthritis, Insect bites and stings, Removes dead skin cells, broken capillaries, Cleansing greasy skin, Corns, Warts, Verrucae, Scars, Strengthens nails, Cuts, Circulation problems, Boosts immune system, Asthma, Bronchitis, Catarrh, Dyspepsia, Sinusitis, Tonsillitis, Chilblains, Mouth sores, Middle ear infections, Depression, Indigestion, Boils, Debility, Fluid retention, Mouth Ulcers, Oily skin, Wounds.

Lemongrass

Note - Top

Type - Citrus

Family - Graminaea

Part - Leaves

Extraction - Distillation

Aroma: Strong, sweet and lemony

Blends well With: Basil, Cedarwood, Coriander, Eucalyptus, Geranium, Jasmine, Lavender, Neroli, Niaouli, Palmarosa, Peppermint, Rosemary, Tea Tree, Thyme, Vetiver, Yarrow.

Contraindications: Sensitive skin, use in low dose, High Blood pressure.

Properties: Antidepressant, Antiseptic, Bactericide, Carminative, Deodorant, Digestive, Diuretic, Fungicide, Galactagogue, Insecticide, Prophylactic, Stimulant, Tonic, Tones the skin, Sedative, Relaxing.

Uses: Exhaustion, Boosting parasympathetic nerves, Digestive complaints, Illness, Loss of appetite, Colitis, Indigestion, Gastro-enteritis, Infectious disease, Respiratory infections, Sore throat, Laryngitis, Fever, Aching muscles, Muscular aches and pains, Tired legs, Jet Lag, Headaches, Fatigue, Insect repellent, Fleas, Aids flow of breast milk, Open pores, Acne, Oily skin, Athlete's foot, Thrush, Digestive disturbances, Stress, Anxiety, Fluid retention, Gastric infections, Gout, Lice, Mental fatigue, Poor Circulation, Rheumatism, Scabies.

Lime

Note - Top

Type - Citrus

Family - Rutaceae

Part - Peel

Extraction - Expression/Distillation

Aroma: Sharp and Tangy

Blends well With: Angelica, Basil, Bergamot, Citronella, Citrus oils, Clary Sage, Geranium, Linden Blossom, Lavender, Neroli, Nutmeg, Palmarosa, Rose, Rosemary, Violet, Ylang Ylang.

Contraindications: Phototoxic, Sensitive skin.

Properties: Antiscorbic, Antiseptic, Antiviral, Apertif, Astringent, Bactericide, Disinfectant, Febrifuge, Haemostatic, Insecticide,

Restorative, Tonic, Stimulating, Activating, Refreshing, Uplifting, Digestive stimulant, Stimulates appetite, Stimulates digestive secretions.

Uses: Apathy, Anxiety, Depression, Fever, Colds, Sore Throat, Flu, Coughs, Chest congestion, Catarrh, Sinusitis, Immune system tonic, Infections, Anorexia, Alcoholism, Rheumatic pain, Greasy skin, Acne, Cuts, Wounds, Anaemia, Brittle nails, Boils, Chilblains, Corns, Herpes, Insect bites, Mouth Ulcers, Spots, Varicose veins, Warts, Arthritis, Cellulitis, High Blood Pressure, Nose bleeds, Obesity, Poor circulation, Rheumatism, Asthma, Throat infections, Bronchitis, Dyspepsia, Headaches, Stress.

Mandarin

Note - Top to Middle

Type - Citrus

Family - Rutacaea

Part - Peel

Extraction - Expression

Aroma: Sweet and tangy

Blends well With: Basil, Bergamot, Black Pepper, Chamomile, Cinnamon, Citrus oils, Clary Sage, Coriander, Cumin, Geranium, Grapefruit, Juniper, Lavender, Lemon, Lime, Marjoram, Nutmeg, Palmarosa, Petitgrain, Rose.

Contraindications: Phototoxic.

Properties: Antispasmodic, Cholagogue, Cytophylactic, Digestive tonic, Emollient, Sedative, Tonic, Uplifting, Stimulates appetite, Stimulates liver, Regulate Metabolic processes, Aids secretion of Bile, Aids breaking down of fats, Calms the Intestines, Revitalising, Skin tonic.

Uses: Depression, Anxiety, Flatulence, Morning sickness, PMT, Stretch Marks and scarring, Digestive weakness, Good for children/ pregnant women and the elderly, Insomnia, Nervousness, Acne, Oily skin, Fluid retention, Obesity, Hiccups, Restlessness, Constipation, Digestive problems.

Marjoram

Note - Middle

Type - Herb

Family - Labiatae

Part - Leaves

Extraction - Distillation

Aroma: Slightly Spicy

Blends well With: Basil, Bergamot, Cedarwood, Chamomile, Cypress, Eucalyptus, Geranium, Lavender, Mandarin, Melissa, Orange, Nutmeg, Peppermint, Rosemary, Rose wood, Tea Tree, Thyme, Ylang Ylang.

Contraindications: May cause drowsiness with prolonged use, Pregnancy.

Properties: Analgesic, Anaphrodisiac, Antiseptic, Antispasmodic, Carminative, Cephalic, Cordial, Digestive, Emmenagogue, Expectorant, Hypotensive, Laxative, Nervine, Restorative, Sedative, Tonic, Vulnerary, Calming, Dilates the arteries and capillaries, Cleanses toxins, Regulates Menstrual cycle, Hypnotic,

Uses: Stress, Anxiety, Psychological trauma, Confronting "issues", Grief, Loneliness, Hyperactivity, Muscular aches and pains, Lower back pain, Rheumatic aches and pains, Swollen joints, Stiff joints, After-sports rub, High Blood pressure, Headache, Migraine, Insomnia, Digestive problems, Stomach cramps, Indigestion, Constipation, Flatulence, Sea Sickness, Chest infections, Colds, Sinusitis, Bronchitis, Asthma, Congestion, Painful periods, Bruises, Sprains, Arthritis, Menstrual problems, Coughs, Colic, Chilblains, Ticks, Lumbago, Strains, Dyspepsia, Leucorrhoea, PMS, Nervous tension, Cramps, Emotional comfort, Flu, Circulation.

Orange

Note - Top

Type - Citrus

Family - Citrus Vulgaris/Aurantium/Sinesis

Part - Peel

Extraction - Expression

Aroma: Zesty and refreshing citrus

Blends well With: Angelica, Black Pepper, Cedarwood, Cinnamon, Citrus oils, Coriander, Clary Sage, Clove, Cumin, Cypress, Frankincense, Geranium, Ginger, Hyssop, Jasmine, Juniper, Lavender, Myrrh, Neroli, Nutmeg, Petitgrain, Rose, Rose wood.

Contraindications: Sensitive skin, Phototoxic.

Properties: Antidepressant, Antiseptic, Antispasmodic, Carminative, Digestive, Febrifuge, Sedative, Stomachic, Tonic, Stimulates bile, Helps digestion of fats, Stimulates Appetite, Aids digestion of Vitamin C, Aids formation of collagen, Lowers cholesterol, Uplifting, Calming.

Uses: Depression, Tension, Stress, Boredom, Lethargy, "Butterflies" in the stomach, Diarrhoea, Constipation, Viral infections, Colds, Bronchitis, Fever, Repairing body tissues, Sore muscles, Rickety bones, Insomnia, Anxiety, Congested skin, Wrinkles, Dermatitis, Skin complaints, Flu, Oily skin, Coughs, Heartburn, Indigestion, Dull complexion, Dyspepsia, Mouth ulcers, Obesity, Palpitations, Water retention, Nervous tension.

Patchouli

Note - Base

Type - Herb

Family - Pogostemon patchouli

Part - Leaves

Extraction - Distillation

Aroma: Earthy and woody, a bit spicy (sort of mouldy)

Blends well With: Angelica, Bergamot, Black pepper, Cedarwood, Clary Sage, Elemi, Frankincense, Galbanum, Geranium, Ginger, Lavender, Lemongrass, Myrrh, Neroli, Orange Pine, Rose, Rose wood, Sandalwood, Ylang Ylang.

Contraindications: Sedative in low doses but Stimulating in high doses, May cause loss of appetite, May cause a headache in some people.

Properties: Grounding, Balancing, Antidepressant, Antipholgistic, Antiseptic, Anti-inflammatory, Aphrodisiac, Astringent, Cicatrisant, Cytophylactic, Deodorant, Diuretic, Febrifuge, Fungicide, Insecticide, Sedative, Tonic, Soothing, Calming, Opens pores, Cell regeneration.

Uses: Lethargy, Clarifying problems, Loose skin, Weight loss, Diarrhoea, Water Retention, Cellulite, Sweating, Insect bites and stings, Snake bites, Scars, Cracked skin, Sores, Wounds, Acne, Eczema, Fungal infections, Scalp disorders, Skin disorders, Athlete's foot, Dry Skin, Oily Skin, Tinea, Excessive Menstrual flow, Dandruff, Dermatitis, Hair care, Impetigo, Wrinkles, Frigidity, Nervous exhaustion, Stress, Anxiety, Bed sores, Depression.

Peppermint

Note - Top

Type - Herb

Family - Labiatae

Part - Leaves and flowering tops

Extraction - Distillation

Aroma: Minty

Blends well With: Basil, Benzoin, Cedarwood, Cypress, Eucalyptus, Jasmine, Lavender, Lemon, lemongrass, Mandarin, Marjoram, Niaouli, Pine, Rosemary.

Contraindications: (Very cooling on the skin - so best used in very small concentration on the skin), Sensitive skin, Pregnancy, Nursing Mothers (can discourage flow of milk), Should not be used with Homoeopathic remedies.

Properties: Analgesic, Antidontalgic, Anaesthetic, Antigalactagogue, Antiphlogistic, Antiseptic, Antispasmodic, Astringent, Carminative, Cephalic, Cholagogue, Cordial, Decongestant, Emmenagogue, Expectorant, Febrifuge, Hepatic, Nervine, Stimulant, Stomachic, Sudorific, Vasoconstrictor, Vermifuge, Encourages sweating, Insecticide, Softens skin, Improves thinking, Uplifting.

Uses: Anger, Hysteria, Nervous trembling, Mental fatigue, Depression, Colds, Mucous, Fever, Respiratory disorders, Dry coughs, Sinus congestion, Asthma, Bronchitis, Cholera, Pneumonia, Tuberculosis, Food poisoning, Vomiting, Diarrhoea, Constipation, Flatulence, Halitosis, Colic, Gall stones, Nausea, Travel sickness, Kidney disorders, Liver disorders, Numbness in the limbs, Shock, Vertigo, Anaemia, Dizziness, Fainting, Headaches, Migraines, Toothache, Aching feet, Rheumatism, Neuralgia, Muscular aches, Scanty menstruation, Painful periods, Mastitis, Dermatitis, Ringworm, Scabies, Pruritus, Itching, Inflammation, Sunburn, Blackheads, Oily hair and skin, Abdominal cramps, Digestive upsets, Flu, Morning Sickness, Shingles, Insect bites, Nervous stress, Indigestion, Acne, Palpitations.

Rosemary

Note - Middle

Type - Herb

Family - Labiatae

Part - Flowering tops and leaves

Extraction - Distillation

Aroma: A strong refreshing herbal scent

Blends well With: Basil, Black Pepper, Cedarwood, Frankincense, Geranium, Ginger, Grapefruit, Hyssop, Lavender, Lemongrass, Lime, Mandarin, Melissa, Myrtle, Orange, Peppermint, Petitgrain, Pine, Tea Tree, Tangerine.

Contraindications: Pregnancy, High Blood pressure, Epilepsy, Not to be taken with Homoeopathic remedies.

Properties: Analgesic, Antidepressant, Antirheumatic, Antiseptic, Antispasmodic, Astringent, Carminative, Cephalic, Cholagogue, Cicatrisant, Cordial, Decongestant, Digestive, Diuretic, Emmenagogue, Hepatic, Hypertensive, Nervine, Resolvent, Stimulant, Stomachic, Sudorific, Tonic, Vulnerary, Invigorating, Strengthens the mind, May help restore speech; hearing and sight, Heart tonic, Cardiac stimulant, Encourages hair growth.

Uses: Memory, Mental Strain, Dullness, Lethargy, Weakness, Mental exhaustion, Headaches, Migraines, Gastric problems, Vertigo, Paralysed limbs, Gout, Rheumatic pains, Tired and overworked muscles, Low blood pressure, Anaemia, Colds, Asthma, Chronic bronchitis, Flu, Hepatitis, Cirrhosis, Gallstones, Jaundice, Blocked bile ducts, Colitis, Dyspepsia, Flatulence, Stomach pains, Menstrual cramps, Scanty periods, Water retention, Cellulite, Obesity, Sagging skin, Skin congestion, Puffy skin, Swelling, Scalp disorders, Dandruff, Infections, Halitosis, Stress, Eyesight, Apathy, Muscle fatigue, Poor circulation, Aches and pains, Acne, Dermatitis, Eczema, Oily hair, Insect repellent, Lice, Seborrhoea, Scabies, Varicose veins, Rheumatism, Whooping cough.

9

Scientific Guidelines for Medicinal Plants

The guidelines for "Good Agricultural Practice of Medicinal and Aromatic Plants" are intended to apply to the growing and primary processing of all such plants as traded and used in the European Union. Hence, they apply to the production of all plant materials used in the food, feed, medicinal, flavouring and perfumery industries.

They also apply to all methods of production including organic production in accordance with the European Union regulations. Those farmers involved in the production of medicinal and aromatic plants should ensure that damage to the existing wildlife habitats is avoided and that efforts are made to maintain and enhance the biodiversity of their farms. It is advisable to regulate wild gathering by a specific guideline.

The present Good Agricultural Practice guidelines provide additional standards for the production and processing of raw materials insofar as they mainly focus on identifying those critical production measures that are necessary in contributing to good product quality.

In this respect, an attempt will be made to minimize insufficient quality by prevention means. The main aim of GAP guidelines is to ensure that the plant raw material meets the demands of the consumer and thus fulfils high quality standards. It is therefore important that they:

- Are produced hygienically, in order to reduce microbiological load to a minimum,
- Are produced with care, so that negative effect on plants in the course of cultivation, processing and storage is limited.

Since medicinal and aromatic plants and their products are exposed to a large number of microbiological and other contaminants in the course of the production process, the main aim of the present guidelines is to provide guidance for producers to reduce this plant raw material contamination to the lowest possible minimum.

All participants in the production process (from primary producers to traders) are required to comply with these guidelines voluntarily and to develop practical measures in order to realize them.

Producers, traders and processors of medicinal and aromatic plants, especially of tea-related and herbal medicinal products, should comply with the GAP guidelines, document their products by a way bill (batch documentation) and demand that their partners also adhere to these requirements.

Guidelines for Good Agricultural Practice

Seeds and Propagation Material

Seeds in use are to be identified botanically, indicating plant variety, cultivar, chemotype and origin. The material used should be 100 % traceable. The same applies to vegetatively propagated parent material. Parent material used in organic production has to be certified as authentically organic.

The parent material should meet the requirements/standards concerning purity and germination (where available: certified seed/ propagation material should be used). The parent material should be as free as possible from pests and diseases in order to guarantee healthy plant growth. If resistant or tolerant species or origins are available, they should be preferred.

The occurrence of plants and plant parts that are not species/ variety identical has to be controlled in the course of the entire production process (cultivation, harvest, drying, packaging). Any impurites have to be eliminated promptly. Plant materials or seeds derived from or comprising genetically modified organsims have to comply with national and European Union regulations.

Cultivation

Depending on the method of cultivation *e.g.* conventional or organic, growers should be allowed to follow different "Standard Operating Procedures" for cultivation (to be elaborated). In general, care should be taken to avoid environmental disturbances. The principles of good crop husbandry must be followed including an appropriate rotation of crops.

Soil and Fertilization

Medicinal and aromatic plants should not be grown in soils that are contaminated by sludge. Furthermore, soils should not be contaminated by heavy metals, residues of plant protection products and any other unnatural chemicals. It should therefore be endeavoured to apply the use of chemical products with as minimum negative effect as possible.

The manure applied should be void of human faeces and be thoroughly composted prior to application. All other fertilizing agents should be applied sparingly and in accordance with the demands of the plant and the particular species (including application between harvests). The use of fertilizers should be in accordance with efforts to minimize leaching.

Irrigation

Irrigation should be minimized as much as possible and only be applied according to the needs of the plant.

Water used for irrigation should comply with national and European Union quality standards and be as free as possible from contaminants, such as faeces, heavy metals, pesticides, herbicides and toxicologically hazardous substances.

Crop Maintenance and Plant Protection

Tillage should be adapted to the growth and requirements of plants. Pesticide and herbicide application should be avoided as far as possible. When necessary, they should be carried out using the minimum effective rates of approved plant protection products. Products for chemical plant protection have to conform with the European Union's maximum residue limits (European Pharmacopoeia, European Directives, Codex Alimentarius). Application and storage of plant protection products have to be in conformity with the regulations of manufacturers and the respective national authorities.

The application should only be carried out by qualified staff using approved equipment. Application should precede the harvest by a period either defined by the buyer or indicated by the producer of the plant protection product.

The use of pesticides and herbicides has to be documented.

All measures regarding nutrient supply and chemical plant protection, should secure the marketability of the product. It is obligatory that the buyer of the batch be informed of the brand, quantity and date of pesticide use in written form.

Harvest

Harvesting should take place when the plants are of best possible quality, according to their different utilizations. Harvesting should preferably take place under the best possible conditions (wet soils, dew, rain or exceptionally high air humidity can be unfavourable). If harvesting is done under wet conditions, extra care should be taken in order to avoid the unfavourable influence of moisture.

Equipment should be kept in clean and technically perfect working condition. Those machine parts, including their housing, that have direct contact with the harvested crop should be regularly cleaned and kept free of oil and other contaminants (including plant left-overs).

Cutting devices of harvesters must be adjusted so that their collection of soil particles is reduced to a minimum.

In the course of harvesting, care should be taken to ensure that no toxic weeds are mixed with the harvested crop.

Damaged and perished plant parts must be promptly sorted out.

All containers used in the harvest must be cleaned and kept free of the remnants of previous crops; containers not in use, must also be preserved in a dry condition, free of pests and inaccessible to mice/rodents as well as livestock and other domestic animals.

The harvested crop should not come into direct contact with the soil. It must be promptly collected and submitted to transport under dry and clean conditions (by use of sacks, baskets, trailers, containers, etc.). Mechanical damage and compacting of the crop that may result in undesirable quality changes must be avoided. In this respect, attention must be paid in ensuring that:

- The sacks are not overfilled,
- The stacking up of sacks does not result in thickening of the crop,
- The harvested crop is transported and kept in containers or bags in such a way that the occurence of heating up is prevented.

Delivery of freshly harvested plant material to the processing facility must occur as quickly as possible in order to prevent its heating up.

The harvested crop must be protected from pests, mice/rodents and domestic animals. Pest control measures should be documented.

Primary Processing

Primary processing after harvest includes such processing steps as washing, freezing, distilling, drying, etc. All these processes, whether for food or medicinal use, must conform to European Union and other national regulations.

On arrival at the processing facility, the harvested crop has to be promptly unloaded and unpacked. Prior to processing the material should not be directly exposed to the sun (except where there is specific need *e.g.* distillation) and it must be protected from rainfall.

Buildings used in the processing of harvested crops must be clean, thoroughly aerated and should not be used for the housing of livestock. Buildings should be so constructed as to provide protection for the harvested crop against birds, insects, rodents as well as domestic animals. In all storage (including packaging) and processing areas, suitable pest control measures, such as baits and electric insect killing machines, must be operated and maintained by professionally qualified staff or contractors.

Processing equipment must be kept clean and be regularly serviced.

In the case of natural open air drying, the crop must be spread out in a thin layer. In order to secure unlimited air circulation, the drying frames must be located at a sufficient distance above the ground. Attempts must be made to achieve uniform drying of the crop so as to avoid mould formation.

When drying with oil, the exhaust fumes should not be reused for drying. Direct drying should not be allowed except with butane, propane, or natural gas.

Except in the case of natural open air drying, the conditions (*e.g.* temperature, duration, etc.) must be selected taking into consideration the type (*e.g.* root, leaf or flower) and active substance content (*e.g.* essential oils and others) of the crude drug to be produced.

Drying conditions should be documented.

Drying directly on the ground with direct exposure to sunlight should be avoided unless it is required for a particular plant.

All material must be sorted or sieved in order to eliminate sub-standard products and foreign bodies. Sieves must be maintained in a clean state and should be serviced regularly.

Clearly marked waste-bins should be kept handy, emptied and cleaned daily.

The dried product should be promptly packaged in order to protect it from and reduce the risk of pest attacks.

Packaging

After the repeated control and eventual elimination of low-quality materials and any foreign bodies, the product should be preferably packaged in new, clean and dry sacks, bags or chests. The label must be clear, permanently fixed and be made of non-toxic material. Information must conform with the European Union and national labelling regulations.

Packaging materials should be stored in a clean and dry place that is free of pests and inaccessible to livestock and other domestic animals. It must be guaranteed that no product contamination occurs as a result of using packaging material, particularly in the case of fibre bags.

Reusable packaging materials should be well cleaned and perfectly dried prior to their usage. It must be guaranteed that no contamination takes place by reusing bags.

Storage and Transport

Packaged dried materials and essential oils should be stored in a dry, well aerated building, in which the daily temperature fluctuations are limited and good aeration is guaranteed. Fresh products (except basil) should be stored between 1°C and 5°C while frozen products should be stored below -18°C (or below -20°C for longer term storage). Essential oil storage must conform to the appropriate chemical storage standards.

As a protection against pests, birds, rodents and domestic animals, the window and door openings should be protected, *e.g.* by wire netting.

It is recommended that the packaged dry crop be stored as follows:

- In buildings with concrete or similar easily cleanable floors,
- On pallets,
- With sufficient distance from the wall,
- With thorough separation from other crops to avoid cross-contamination
- And organic products must be stored separately.

In case of bulk delivery, it is important to ensure that the transportation conditions are dry. Furthermore, it is highly advisable to use aerated containers in order to reduce the risk of mould formation or fermentation. As a supplement, the use of other sufficiently aerated transport vehicles and other aerated facilities is recommended. Essential oil transportation must conform to the relevant and applicable regulations. In this respect, national and European Union regulations on transport have to be adhered to.

Fumigation against pest attack should only be carried out on necessity and must exclusively be administered by licensed personnel. Only registered chemicals should be used. Any fumigation against pest attack should be reported in the documentation.

For fumigation of warehouses, only permitted substances should be used in accordance with European Union or national regulations.

Whenever frozen storage or saturated steam is used for pest control, the humidity of the stored material must be controlled after each treatment.

Equipment

Equipment used in plant cultivation and processing should be easy to clean, in order to eliminate the risk of contamination.

All machinery should be mounted in an easily accessible position. They must be well serviced and cleaned regularly. Fertilizer and pesticide application machinery must be regularly calibrated.

Preferably, non-wooden equipment should be used unless tradition requires the use of wooden material. Whenever wooden equipment is used (*e.g.* pallets, racks, bunkers, etc.), care should be taken to avoid its direct contact with chemical and other contaminated/infected materials, so as to prevent the impurity of the plant material.

Personnel and Facilities

Personnel should receive adequate botanical education before performing tasks that require this knowledge. All processing procedures should fully conform with both EU-guidelines on food hygiene and the general food hygienic principles of Codex Alimentarius as well as the European directive on Good Manufacturing Practice (GMP). The personnel entrusted with the handling of plant material are required to exhibit a high degree of personal hygiene (including those personnel working in the fields) and should have received adequate training in their hygienic responsibilities.

The buildings where the plant processing is carried out have to be equipped with changing facilities as well as toilets and hand washing facilities, with respect to the relevant regulations. Persons suffering from known infectious diseases transmittable via food, including diarrhoea, or being transmitters of such diseases, must be denied entry into areas where they may come into contact with the plant material, as per relevant regulations.

Persons with open wounds, inflammations and skin infections should be kept away from those areas where plant processing takes place, or have to wear appropriate protective clothing or gloves, until their complete recuperation.

Personnel should be protected from contact with toxic or potentially allergenic plant materials by means of adequate protective clothing.

The welfare of all staff involved in the growing and processing of medicinal and aromatic plant products should be ensured.

Documentation

All parent materials and processing steps, including the location of cultivation, have to be documented. Field records showing previous cropping and used inputs should be maintained by all growers.

All batches from coherent areas should be clearly and unmistakably labelled (*e.g.* by the application of a batch number). This should take place as early as possible.

Batches from different cultivation areas may be mixed only if it is guaranteed that the mixture in itself will be homogenous. Such mixing procedures should also be documented.

It is essential to document the type, quantity and date of harvest of the crop, as well as the chemicals and other substances (*e.g.* fertilizers, pesticides and herbicides, growth regulators, etc.) used during production. The application of fumigation agents such as phosphin must be entered into batch documentation.

All processes and procedures that could bear any impact on the quality of the product must be entered into the batch documentation.

All agreements (production guidelines, contracts, etc.) between the producer and the buyer should be laid down in written form.

It should be documented in a way bill (batch documentation) that cultivation, harvesting and production have been performed in accordance with the GAP guidelines. Minimal information included in the way bill should cover the geographical definition of growth place, the country of origin and the responsible producer.

Audit results should be documented in an audit report (*i.e.* copies of all documents, reference index, audit reports, analysis results) and be stored for a minimum of 10 years.

Any special circumstances during the growth period which may influence the chemical composition, *e.g.* extreme weather conditions or pests - particularly in the harvest period - must be documented.

Education

It is highly advisable to educate all personnel dealing with the crop or those engaged in the management of production, in production techniques as well as the appropriate use of herbicides and pesticides.

Quality Guarantee

Consultation between producers and buyers of medicinal and aromatic plants, with regard to quality questions, *e.g.* active principles and other characteristic ingredients, optical and sensoric properties, limited germ numbers, plant protection chemical residues and heavy metals, must be based on nternationally recognized or national specifications and should be laid down in written form.

Medicinal Plants

The American public has a longtime love affair with herbs. A few years ago everyone was grinding basil for pesto, stuffing rosemary in their vinegar bottles, and putting French tarragon sprigs on their chicken. Then they moved on to herbal wreathes, soaps, and potpourris. Now companies across the nation indicate that herbal medicines and aromatherapy are the fastest growing sales categories. The general feeling in the industry is that the public's interest in herbs has not reached its peak and healthy growth is expected in the future.

So what can that mean for you as a propagator? Retail nurseries report that gardeners are very interested in growing their own medicinal herb plants. People are in the market for shade loving perennials, such as ginseng and goldenseal; sun-loving plants, such as purple coneflower; and shrubs and trees, such as witch hazel and ginkgo.

The companies producing medicinal herb products, such as the capsules and tinctures you find in health food stores and pharmacies, have traditionally purchased most of their raw herbs from India, China, and the Eastern European countries. As the medicinal herb industry has matured, however, the companies buying these herbs have instituted new quality control efforts that include testing for active compounds, purity, and bacterial contamination. Many companies have found that foreign imports do not meet their new high standards. This has opened up a market for American growers to produce large acreages of medicinal herbs. This, in turn, has created a market for the seeds and plants to produce these herbs. That is where you as a propagator can get involved with herbs. So let me tell you about some of the plants that are currently in demand and some of what is known about producing these plants.

There are several important points to keep in mind before you jump into medicinal plant production. One is that you are working with "new crops". Thus, your county extension office will not have information packed bulletins on how to grow most of these plants. I encourage you to read everything you can about the herb plants we do know something about, and about plants that are similar to what you want to grow. Talk to the people at your local botanical garden and to herb enthusiasts. Mostly, you must rely on your own experience, horticultural knowledge, and "gut instincts." Keep careful notes on everything you do and the results you obtain. You could quickly find yourself the expert on propagation of one of these little known medicinal herbs.

Second, be very careful about where you buy your seeds and planting stock. This is a new industry and quality standards aren't always what they should be. Check your sources out very carefully. If you decide to collect seeds and planting stock from wild populations, be certain to check on local regulations and the status of plants before you head out. Some of the more popular medicinal herbs are endangered or protected species in some states.

For any of the shade-loving plants, of course, you must provide shade. Commercial herb growers use natural shade or artificial shade provided by wood lath or polypropylene shade cloth. When using artificial shade, I recommend tall structures, about 8-9 feet tall, with two open ends for good ventilation.

What are some of the plants in demand? Right now, one of the plants in greatest demand is goldenseal; Hydrastis canadensis, a shade loving, herbaceous perennial. It is used for many purposes including as a treatment for AIDS, cancer, various digestive disorders, and to boost the immune system. It is native to North America and used to grow in abundance throughout the Appalachian region. Unfortunately, native populations have been seriously reduced by over collection and it is now an endangered species in North Carolina and under new regulations for international trade.

Goldenseal has a rhizome by which it spreads rapidly and is easily propagated. In fall or spring, these rhizomes can be cut into pieces. Try to include an obvious bud on each piece, but research indicates that it is even more important to have good roots. Goldenseal produces a raspberry like fruit, which is full of seeds. These seed require special handling, such as they cannot be allowed to dry out. Most growers find they get the best germination when the seed are sown immediately after they are extracted from the fruit. This, of course, is not real practical for the propagator.

Studies are underway at North Carolina Statue University looking at seed extraction methods, seed disinfection treatments, seed storage temperature regimes, and planting dates. So far, the best germination has been obtained from seed that had been held at 70 degrees F prior to sowing in late August or late October.

Studies are also being conducted on soil fertility, mulch, and spacing for goldenseal. For the fertility studies, goldenseal was grown for three years in forest soil amended to provide four levels of lime, phosphorus, and nitrogen. Emergence the first year was greatly influenced by lime application. Emergence was very low where the soil pH was lower than 4.8 or higher than 6.7. If the plants made it through that first year at the undesirable pH levels, they exhibited 90% or better germination the following years. Any supplemental nitrogen, reduced emergence that first year, but had little effect in the following years. The response to phosphate, supplied as superphosphate, was much the same. As with emergence, the plants were tallest around pH 5.5 and 6.5.

Goldenseal roots and rhizomes are the plant parts of interest to the propagators and the botanical ingredient companies. There was a dramatic response in growth of these parts to soil pH. The bottom line is, remember to keep pH around 5.5 to 6.5. These numbers will be refined as further research is conducted. Fresh root weight was inversely related to increasing nitrogen rate during the first year of growth. There were no significant responses to nitrogen rate after the second or third year, however. The response to phosphate additions was very similar.

Studies are now underway looking at spacing effects. Plants emerged first in the closest spacings of 2x2 inches and 4x4 inches. Mulch studies were also initiated in 1997. Plants grew best the first season with bark mulches, pine needles, and sawdust. Straw mulch resulted in poor growth and severe slug damage. Shade requirements are also being examined with plants growing under polypropylene shade providing 30%, 47%, 63%, 80% shade and wood lath.

So, as you can see—there was a lot we didn't know, and still don't know, about growing goldenseal. However, for those willing to try and work out the cultivation methods, there are good markets for goldenseal seedlings and mature plants for planting stock. There also is a limited, but growing market, for goldenseal plants for retail sales.

American ginseng (Panax quinquefolium) is another medicinal herb that has a long history of cultivation in North America. You will have no problem finding information on ginseng seed handling and plant growing. Ginseng seed requires very special handling and a stratification period of about 18 months. Like goldenseal, ginseng seed must never be allowed to dry out. There are a large number of suppliers of ginseng seed and roots for commercial growers. However, there are few nurseries catering to the home gardeners. A few enterprising growers have tapped into this lucrative hobbyist market. These plants sell for $12 to $25 a piece.

Other shade requiring medicinal plants in demand include black cohosh (Cimicufuga racemosa)—a perennial shrub, fairly easy to propagate by root division or seed. The roots of this plant are used to treat various feminine disorders. Mayapple (Podophyllum peltatum) is another woodland plant in demand because of its cancer fighting properties.

It is propagated primarily by dividing the roots in the spring because the seeds can be hard to germinate. The plant thrives, however, under cultivation. Bloodroot (Sanguinaria canadensis) has a pretty, early spring flower which makes this plant especially attractive to the retail market. Bloodroot can be easily propagated by seed or root division. Bloodroot is used in a popular toothpaste because of its antiplaque properties. It also has other antibacterial properties. It can be quite toxic, however, at low doses!

Chamelirium luteum goes by many common names, including fairy wand, star grub root, Helonias, and devil's bit. The plant is an erect, fleshy, perennial with a rosette of leaves at the base. Male and female flowers are borne on separate plants. This plant has not really been cultivated and we know little about its requirements. However, demand for this plant is increasing. The roots are used to treat a variety of feminine disorders. Any herb that can be of use to the aging baby boomers, especially to help them deal with problems related to menopause, prostate, poor memory, and low energy, will have a strong market!

A sunloving, field grown plant that does well for plant sales include all the different species of Echinacea, including angustifolia, purpurea, and pallida. The most common use for Echinacea right now is to help boost the immune system to prevent catching colds and the flu. Skullcap (Scutellaria lateriflora) is in demand as transplants for a field crop. It is easily grown from seeds. Skullcap is used as a sedative and to treat headaches. Calendula (Calendula officinalia) is a very attractive herb, commonly used in home herb gardens but also grown in large acreages. It is easily grown from seed. Calendula is valued for its wound healing properties.

If you decide that you want to get into the medicinal herb plant business, you need to know what is happening in the industry. My first recommendation is that you subscribe to some of the more popular herb publications. The demand for various herbs can change rapidly, and it is important to keep abreast of what is happening in the industry. For information on the industry in general, with emphasis on retail, look at publications such as the "Business of Herbs" and the "Herbal Connection". Information of this sort on the wholesale and field grown medicinals is harder to come by, but "HerbalGram" and the trade publications for natural foods are good places to start. "ATTRAnews," the newsletter from the federally funded programme called Appropriate Technology Transfer for Rural Areas, is a publication that is useful to anyone looking at commercial production of alternative crops.

Some of the companies buying herbs and contract growing also can provide some useful production information. And some seed companies who sell large volumes of herb seed provide technical assistance. For example, Johnny's Selected Seeds hired medicinal plant experts as consultants to help with their herb line and Richter's Herb Seeds holds a growers conference every year. I also recommend that you join local and national herb associations. For example, the International Herb Association provides good networking opportunities, holds a national conference each year, and has regional chapters that can be a good source of information.

Currently the market is strong for medicinal herbs so many growers are jumping into it. But the ones that stay with it and find it profitable will be the ones who do their homework, grow a quality plant, and provide the service and prices the market demands.

Medicinal Plants of Meru

Since ancient times, Meru people in Kenya have largely depended on their local plants for medicines, food, fuels, material and environmental uses including hedges, shades, and land markers. Meru bee plants are used in beehive smoking and bee attractors, while other plants for social use include masticatories and those used in games and in smoking. Traditional cosmetic and aromatic plants are used for Meru socio-cultural puroposes. In the ethnobotanical study conducted in Meru (Kenya) in 2002 and 2003 was found that Meru people know and use over 635 different species of plants for the various needs listed above. Of the total number of 635, not less than 411 were medicinal plants, while a few were cosmetic and aromatic plants. The study has documented the indigenous plant classification, local use and management practices, and related illnesses.

Medicinal Plants in Health

Various trees, shrubs, herbs, climbers and creepers/climbers, grass, succulents, bulbs and rhizomes are mainly used to prepare traditional medicines. The specific parts of the plant that make the medicines include bark, roots (tuber), leaves and buds. The preparation of herbal medicine varies for 'external' and' internal' illnesses.

A large number of perceived diseases have been reported to be mainly treated through the application of specific traditional herbal medicine. These include: head, tooth and gum aches, sore eyes, growths on the face, common colds, and dandruffs, brain fatigue, tonsillitis, mumps, mouth sores and aching ears. Also, neck stiffness, asthma, coughs, chest pains, pneumonia, tuberculosis, and whooping cough belong to this category of diseases.

Stomach ailments treated by herbal medicine include acidity, ulcers, worms, heartburn, diarrhoea, constipation, amoeba, poisons, anthrax, nausea and cholera. Backaches, joint aches, swollen legs, sprains and rheumatism are also among the diseases treated by traditional plant-based medicines. Special herbal remedies are reported for malaria and other kinds of fevers, rashes, boils, measles, itching, scurvy, skin growths, dry skin and ringworms. Similarly, venereal diseases, fresh cuts, wounds from poisoned arrows, burns, anaemia, diabetes and high blood pressure are mostly treated by traditional herbal medicines.

Antenatal herbal treatment include birth control; abortifecients and menstrual overflow while post-natal treatments include problems related to afterbirths and excessive bleeding.

Medicinal Plants in Culture

Plants in Meru have been part of the local people's culture for as long as they have known them. There are particular trees that are believed to give long life - the Mugumo tree; those that give healing by just singing praises to them (Mutuntu) and those that bring prosperity and health to couples and families (Kirao).

Cultural diversity is also manifest in Meru people views on herbal medicines. The differential perceptions are influenced by a variety of factors. These include socio-demographic *e.g.* ethnicity, religion; psyco-social *e.g.* knowledge, attitudes and beliefs; and enabling factors *e.g.* socio-economic status and profession of respondents.

Cosmetic and Aromatic Plants in Health and Culture

Although fewer people tend to use plant-based products nowadays due to the already available cosmetic and aromatic products found at modern chemists, there is recently a reorientation towards the use of cosmetic plants. Some parts of Kenya still use wild plants for their cosmetic products, but the Meru people only knows these specific plants from the past.

The main reported uses then include traditional perfumes for circumcised girls and women. Among young children, particular roots of cosmetic plants are crushed and the powder mixed with saliva and applied on the body of a child, also to cure colds and coughs. An example of such a plant is the Ndago plant.

Medicinal Plants—Industrial Profiles

Which well known plant never touches the ground, does not follow a 12-month vegetation cycle, stays green year round, blooms in the winter, and sells predominately in December? The answer, of course, is mistletoe. Current research questions the efficacy of this highly controversial plant. "Rejected by clinical oncologists but used by practitioners and cancer patients, it is applied as a remedy to treat a broad spectrum of different diseases, such as epilepsy, diabetes, hypertension, arthrosis, hepatitis, HIV infection, labour pains, and cancer." Now a new book on the therapeutic uses and industrial applications of Viscum has been published. Rarely mentioned in traditional books on herbal medicine mistletoe research remains questionable. Because the "extracts may exert different effects than purified substances" individual immune responses vary greatly. There is a great need for further investigation. Currently no clinical data are available documenting long-term therapy treatment.

Whether the extracts of Viscum become a major player in the fight against cancer, this book makes for provocative reading. Researchers in pharmacognosy and phystochemistry will find this book documented with references to hundreds of research articles invaluable. While most articles feature the European point of view, there is one article from Japan and another from South Africa. Although none of the articles originate from the United States, American research is well represented in the bibliographies, as are traditional Chinese uses for this herb.

Responding to the increased demand for information on medicinal and aromatic plants this fascinating look at this highly controversial plant is the 16th volume in the series Medicinal and Aromatic Plants

Industrial Profiles. This series brings together articles scattered through a variety of sources, is international in scope, and includes many little known sources. Each article is well documented and is introduced by the editor Arndt Bussing, head of the Department of Applied Immunology at the foundation Krebsforschung Herdecke, University of Witten/Herdecke. Mistletoe: the Genus Viscum is illustrated with colour photographs of the plant, line drawings, and chemical structures. With few exceptions most references are current, dating from 1990 to 2000.

While this book would be of interest to scientists in botany, agriculture, and phytochemistry, those working in alternative medicine, pharmaceutical, and health sciences should find it of greater interest. Recommended for academic and research libraries.

New Support for Medicinal Plants

Aspirin, ginseng tea, Aloe lotions and gels are simple healthcare remedies used in many parts of the world to treat a variety of illnesses. They all also have a common beginning: wild plants. Aspirin, for instance, owes its origins to the willow tree.

Hippocrates used powder extracted from willow bark to treat pain and reduce fever in the fifth century BC. Over 2000 years later, the chemical compound salicin was isolated from willow bark, the precursor to the active compound found in modern aspirin. Today, it is estimated that more than 80 billion aspirin tablets are used each year in the USA alone. Unlike aspirin, ginseng tea, used to improve vital energy and slow aging, is still produced from the raw product: ginseng roots. However, much of the ginseng in trade is cultivated, as is all of the aloe vera gel found in skin-care products and health drinks. Less well known are the tens and perhaps hundreds of thousands of other medicinal plant species used around the world, some of which, like willow trees, have been recognised for their healing properties for thousands of years.

These plants form the basis of health care systems throughout much of the developing world, a source of new compounds on which to base new pharmaceutical products, and a major component of the burgeoning markets for herbal health care remedies and natural products. They also provide a source of income for growers, traders, collectors and manufacturers of plant-based medicines.

In this dual role as a source of healthcare and income, medicinal plants make an important contribution to the larger development process. Unfortunately, however, some plant species of medicinal value are already in short supply, which poses a threat to human welfare and the wild species themselves. Immediate actions are therefore required to ensure harvest and trade of medicinal plant species are conducted sustainably.

TRAFFIC has been working on the medicinal plant trade for many years, including collaboration with important partners such as the IUCN/SSC Medicinal Plant Specialist Group.

Germany's Federal Ministry for Economic Cooperation and Development (Bundesministerium fur wirtschaftliche Zusammenarbeit, more commonly known as BMZ) is supporting a suite of new actions by TRAFFIC. Through a Funds in Trust Agreement with WWF International, BMZ has pledged DM1 million (US$560 000) to the TRAFFIC Network's medicinal plant work during 1998-2000. Much of the funding is directed toward regionally focused work in East Asia, the Indian subcontinent and South America.

In addition, BMZ is supporting efforts to build and/or strengthen links with others working on medicinal plant trade issues. It aims to enhance the communication of TRAFFIC's research results to those who need it, from traditional medicine communities to policy makers.

The regional work to be undertaken through BMZ support is as diverse as the medicinal plant trade itself, with each region home to different, but in the case of India and East Asia, overlapping, medicinal traditions.

East Asia

The major traditional medicine systems in this region are derived from traditional Chinese medicine (TCM), which has a history of several thousand years and is well-documented.

Although international attention has been drawn to the use of animal species such as rhinoceros and Tiger in TCM, animal ingredients actually make up less than 20% of those used in TCM. TCM is largely a plant-based medicinal system. It is believed that as many as 1000 plant species are in use for TCM in China and that about 80 per cent of them are taken from the wild. An initial TRAFFIC East Asia study in Hong Kong in 1995/96 revealed a massive trade involving a number of threatened plant species: of the 388 plant species listed in the China Plant Red Data Book, 69 were found to be used in TCM, and 22 of these were in common use.

Through work supported by BMZ, TRAFFIC East Asia seeks to promote the security and conservation of wild plant resources used in traditional medicine in East Asia to support basic health care needs. TRAFFIC East Asia will convene a series of meetings to promote dialogue and understanding of the links between conservation and medicinal security issues among producers, regulators and medicinal communities.

Other activities planned include the compilation of an overview of the nature of and trends in medicinal plant harvest, propagation and trade in East Asia; identifying species at risk; and reviewing national legislation and policies relevant to production, supply and conservation of medicinal plants. Work will also be undertaken to help build capacity within governments and TCM communities to implement any needed trade controls.

The Indian subcontinent

The Himalayas provide more than a meeting point for China and the Indian subcontinent. They also provide a huge variety and volume of medicinal plants used in TCM, Tibetan medicine, and Ayurved. Like TCM, Ayurved has a long history (over 3000 years), is well-documented, and is widely used both locally and in Asian communities established in other parts of the world. There are also a large number of localised traditional medicine systems.

Harvest of native plants to supply India's own healthcare needs and a growing export market provides an important source of income for local communities. However, demand for plant materials is huge and growing, while many of the region's medicinal plant resources are dwindling. Recent reviews applying IUCN threat criteria found that a significant number of medicinal plant species are threatened or endangered within India. Large quantities of raw medicinal plant materials are also imported into India from neighbouring countries and exported to overseas markets.

A number of organizations within India are concerned with maintaining India's traditional medicine systems. In addition, there is a widespread development network, an established pharmaceutical industry and a wealth of botanical experts in the country. Until now, however, there has been little effort to document the volume and impact of national or international trade in India's medicinal plants.

TRAFFIC India is already well on its way toward establishing working relationships with key stake-holders such as those above and government personnel and toward collecting information required if plants in trade are to be managed on a sustainable basis. A key challenge within India will be to inform these and other stake-holders of the nature and impact of the medicinal plant trade, and to facilitate their co-operation in developing and implementing measures that will secure the conservation of medicinal plants. Original research will be combined with meetings and workshops to discuss findings and develop workable solutions to any problems identified.

South America

In contrast to East Asia and India, South America's traditional medicine practices are far more localised, with neither the geographic extent or the written texts that characterise TCM and Ayurved. As a result, an important first step is to study and describe the medicinal plant trade in the region, to identify commonly used species, and to begin the process of identifying whether any species are at risk.

Preliminary research in 1996 and 1997 by TRAFFIC International's South America Plants Officer confirmed that, as in Asia, medicinal plants are important components of local health care, are in some cases used unsustainably, and are exported as well as used locally.

BMZ funding is supporting the completion of a regional overview of the medicinal plant trade as well as more detailed studies of individual species in trade, existing trade controls, and the potential role of regional and international treaties in this region. Two species already identified for further research are Cat's Claw Uncaria tomentosa and Dragon's Blood Croton lechleri.

The results of this research will be shared with governments, traditional healthcare associations and relevant agencies to further raise the profile of medicinal plant trade issues and stimulate discussion and actions necessary to ensure more effective management.

Medicinal Plants and CITES

The Challenge of Sustainable Harvest

Plants of medicinal value originate from all over the world. Harvest occurs for both local and international use. A growing problem, which is compounded by global population growth and increased demand for raw materials, is over-exploitation of these important resources. The vast majority of medicinal plants used in traditional medical systems and natural products industries are collected from the wild. While propagation occurs for a few species such as American Ginseng Panax quinquefolius and Kuth Saussurea costus, propagation is generally limited due to the expense and time periods required for plants to reach salable size. Harvest and trade can be an important source of income for rural peoples. Regrettably, unsustainable harvest is an increasingly common occurrence that threatens not only the future of these species, but also the livelihoods and health status of people throughout the world.

The Current Situation

At present an estimated 230 plant species included in CITES are traded for medicinal purposes, or are likely to be traded in the future. Fourteen of these were specifically included in the Appendices expressly because of the need to regulate trade in medicinal material. A further six species, Adonis vernalis, Camptotheca acuminata, Cistanche deserticola, Harpagophytum procumbens, Harpagophytum zeyheri and Panax ginseng, are proposed for inclusion in Appendix II at COP11.

The Parties' implementation of the Convention for the species currently included in the Appendices is poor, and is in many cases non-existent. This is surprising, given the importance of these species to the health care of the peoples of these nations, to the livelihoods of the millions who derive income from medicinal plant harvest and trade, and to the industries and consumers who benefit from these resources.

Indeed, some Parties have made notable efforts to implement CITES for medicinal plants, such as Germany with the publication of the Checklist of Medicinal and Aromatic Plants and their Trade Names covered by CITES and EU Regulation 2307/97 and Identification Training for Medicinal and Aromatic Plants covered by CITES and EU Regulation 2307/97, and USA with the establishment of a ginseng management programme to monitor wild populations and regulate harvest and commerce.

The Plants Committee has examined the trade through the Medicinal Plant Significant Trade Study, and additional significant plant trade surveys have been undertaken by TRAFFIC. However, these efforts are few in number and although Parties may have taken domestic measures to conserve native plant populations, in general, implementation of the Convention for medicinal plants is unsatisfactory.

Over-exploitation of medicinal plants, and lack of attention to illegal and unsustainable trade, is having serious repercussions around the world. As the global population expands, the demand for medicine increases correspondingly. International trade in medicinal and aromatic plants is already massive, with for example, annual global market imports totalling 440,000 metric tonnes in 1996, at a value of USD 1.3 billion. The medicinal plant trade can make a considerable contribution to the economies of producer countries.

Searching for Solutions

As medicinal plant populations decline, governments, natural resource managers, local communities and a variety of other stakeholders have attempted to address the problem in order to conserve medicinal resources. Efforts have been varied.

For example, these initiatives include: promotion of sustainable harvesting techniques; improvement of marketing and trade opportunities to ensure better management; training in quality control and packaging to reduce wastage and derive higher earnings; increased regulation and control at local, national and international levels; and propagation of medicinal plant species. Increasingly, these efforts are designed taking into consideration the complexity of the medicinal plant issue and the realization that demand will not decrease, hence solutions must factor in the aspect of a continuously increasing trade.

The market for medicinal plants is dynamic and complex. Demand for new natural products, and growing interest in traditional medical systems around the world, means that greater volumes of plant material are required to satisfy the market. Medicinal plants are traded in a variety of forms, for example: raw material such as leaves, bark, and roots, unprocessed broken material such as chips or slices, derivatives and semi-processed or manufactured material such as powders, extracts, tonics, pills, teas and other products, as well as finished pharmaceutical products.

Trade occurs far from the area of production, and processing often takes place in countries other than those where the plants originate. For example, approximately 120,000 metric tonnes of medicinal and aromatic plants were imported into Europe in 1996, and this material came from over 120 countries, 60% of which were African and Asian nations.

During this same year, Europe exported about 70,000 metric tonnes of plant material. Understanding the dynamics of the trade in medicinals can be difficult, and although overall volumes and values may be known, species-specific information is not at all comprehensive. The dearth of such data is one reason for the lack of attention countries have directed toward the medicinal plant resource.

Insufficient knowledge about the scope of the trade, in particular at the range state level, has resulted in ineffective implementation of existing regulations and initiatives in producing countries, to the detriment of the resource and those who will depend on it in the years to come. Perhaps most important is that there is little appreciation for the increasingly critical imbalance between supply and demand.

Can CITES Help?

CITES addresses only a small portion of the medicinal plant issue, but it has the potential to help achieve sustainable and legal trade in medicinal resources. Specifically, CITES can do the following:

- identify trends at the species level and on the global level, to highlight critical issues and address them;
- help to make the trade more transparent, by documenting the species and parts in trade, and volumes traded, because this information is necessary for designing and implementing sustainable management plans;
- stimulate cooperation between importing and exporting countries to ensure that trade is conducted in a legal and sustainable manner;

So What is the Problem?

When the trade is so large, and the implications of unsustainable harvest and trade so serious, why is implementation of CITES for medicinal plants unsatisfactory? Some possible explanations include the following:

- Parties are unaware that CITES-listed medicinal plants are in international trade, because of insufficient knowledge about the parts and derivatives that aretraded.
- Regulatory officials responsible for inspection of plant material are unfamiliar with CITES and do not know that certain traded plant species require CITES documentation.
- Plants are frequently afforded less attention than animal species.
- The variety of parts and derivatives in trade is numerous, and the annotations for CITES- listed medicinal plants vary from species to species, therefore it may be difficult for inspectors to keep track of what should be regulated.
- The majority of trade in medicinal and aromatic plants is not regulated by CITES, and therefore CITES-listed species may be accorded a low priority.

Will these Efforts help Medicinal Plants?

At present the sustainability of the medicinal plant trade is on precarious ground. Through CITES, there is an opportunity to promote sustainable harvest and trade. While regulation is only part of the solution, it is an integral part of a strategy that must entail efforts from many sectors and many directions.

Because of the importance of a sustainable trade in medicinal plants, it is entirely appropriate for CITES to collaborate and indeed advise other global bodies that deal directly with human health issues, on the conservation aspect of this problem. With regard to medicinal plants, the objectives of CITES and the CBD are complementary.

Demand for Medicinal Plants in KwaZulu-Natal

Medicinal plants are used extensively by the black population throughout KwaZulu-Natal (KZN), and indeed South Africa. An active trade and market in medicinal plant material occurs in KwaZulu-Natal as a result of:

- a fragmented distribution of medicinal plant species due to the varied land use, topography and climate in the region,
- large spatial concentrations of consumer demand as more than half the province's population is concentrated in two urban centres,
- local extinctions of plants and the need to access more remote plant populations due to intensive harvesting, and
- various socio-economic factors (including culture and accessible health care) which maintain and stimulate demand.

In KwaZulu-Natal over 400 plants are actively traded [Cunningham 1988; McKean 1996] with an estimated trade volume of some 4 300 tonnes per year. As the market is large and dynamic, several market segments have developed in response to the users' demands and the supply of plants. The extent of trade and the market segments are discussed in the following sections.

Quantity and Trends in Consumer Demand

A survey of several Durban City medical clinics indicated that 53% of the black clinic patients made use of indigenous healers at least once during the previous year, with an average of 3.34 visits in the previous year.

The term indigenous healers or medicine is preferred by the author to the more conventional term traditional healers or medicine. The healing practices used, while originating in African culture, are not purely traditional. The practice is dynamic, addressing new illnesses like AIDS, and adopts new technologies and new medicines.

The healers also deal with all manner of urban social problems, which are not traditional. The term indigenous medicine may therefore be more appropriate as it relates to the practices of healing and divination within an indigenous but modem African culture which may not necessarily be traditional.

While respondents asked whether they visited a indigenous healer in the previous year, the survey omitted to ask respondents whether they purchased indigenous medicine without a healer's prescription. As a consequence, the above estimate of indigenous medicine use may be an underestimate. An indication of the extent of self-medication can be obtained from the expressed preference for hygienically packed medicines.

Eighty-four percent of the black clinic patients indicated that they would prefer more hygienically packed indigenous medicines, indicating that 31% of the sample may not have visited an indigenous healer in 1995 but probably make use of indigenous medicine.

Assuming that the clinic sample is representative of the black population in Durban, the total number of users is estimated to be around 2 million people [that is 84% of 2.442 million blacks resident in Durban (Durban Metropolitan Government 19.96)]. This estimate can be considered a conservative minimum level, as 6% of the indigenous healers patients indicated that they did not make use of western clinics, and therefore using the clinic patients as a proxy for the Durban black population may be a slight underestimate.

The Quantity of Plant Material Reported to be Consumed

Basing the quantity demanded on the reported number of times which the population sample visits indigenous healers (3.34 times per annum), it is estimated that between 900 tonnes and 1 500 tonnes of plant material were consumed in Durban in 1995. As most of the estimates are based on relatively imprecise data regarding the numbers of market participants, the broad estimates will generally be rounded off to make for easier reading. However, detailed measurements made within the survey will be retained for descriptive accuracy and where calculations are made.

This estimate assumes that the indigenous medicine users not visiting indigenous healers, that is the remaining 31% of users, would use indigenous medicine at the same rate at which healers are visited (3.34 times a year). The estimate of 900 tonnes is derived if 53% of the population is assumed to use medicine at an average mass of 83.3g per use (the average mass of plant material prescribed by healers and shop traders) and the remaining consumers use the medicine at 216.5g per use (the average mass of plant material traded in the street markets).

The estimate of 1 500 tonnes is obtained if all 84% of the black population uses medicine at 216.5g per use. This last estimate is likely to be more realistic given the extent of wastage associated with the poor storage of plant material.

The Quantity of Plant Material Reported to be Traded

The shop traders are estimated to trade 340 tonnes per year, with approximately 50% reported to be purchased directly from plant gatherers (170 tonnes) and the remaining 50% purchased from street traders (170 tonnes).

The quantities traded by the healers was problematic to estimate due to the unknown number of healers in the survey area. Various reports [Holdstock 1978; Gelfand 1985; Cunningham 1988] had been made regarding the ratio of indigenous healers to the population but these have not been used for a number of reasons.

Firstly, the mode of operation differs considerably between healers, with some healers being extremely busy whilst others may only practise part-time. Secondly, the various healers' organizations have provided a rough indications of the number of healers which are registered with them, but it has not been possible to obtain membership lists from these groups (in addition, the membership makes no differentiation regarding the nature of the practitioners operation, that is, whether it is either full-time or part-time).

Due to the above problems, estimates are based on the visitation frequency reported by the clinic patients and the average number of patients, which the indigenous healers indicate they may treat per day. Following this approach an estimated 936 tonnes of medicinal plants are traded per year by indigenous healers in Durban.

The indigenous healers reported that they purchase plant material from two sources, approximately 50% from the street markets and 50% directly from plant gatherers. This suggests that approximately 468 tonnes are bought from each group.

This estimate uses an average mass of 216.5 g for a prescribed medicine dosage. This mass is the average mass of street traded products and is believed to be a more appropriate mass than the mass dispensed in traders shops and by healers (an average 83.3 g) for estimating the quantity used by healers. The traded products are generally unprocessed and poorly packaged, with a significant proportion (not quantifiable at present) of the plant material spoiled in the transportation and storage process. The author therefore considers the larger street traded mass to be a more accurate reflection of the quantities which healers will utilise but not necessarily dispense.

Of the plant mass traded in the street markets, an estimated 20% of the trade or 220 tonnes, is reported to be traded between the street traders themselves and not to consumers, healers or shop traders. This implies that of the 1 100 tonnes traded in the market as a whole, only 880 tonnes of plant material is the physical stock of plants entering the street markets in a year.

By summing the quantities sold by the various market players to end consumers, it is estimated that some 1 500 tonnes of plant material were sold in the Durban markets in 1996.

This estimate of 1 500 tonnes traded in the market approximates the estimate derived in 4.1.1 regarding the maximum quantity of plant material which may be consumed in Durban given the consumption rates of the clinic patients interviewed. While the two approaches are not entirely independent as both models use a common number in estimating the use associated with healers, there is sufficiently different information in the other sectors which gives confidence to the similarity of estimates.

Potential Provincial and National Consumption Levels

Several studies in different geographic localities throughout South Africa and in different cultural groups have shown that between 58 and 100% of the black population uses indigenous medicine. Holdstock [1978] estimated that between 80 and 85% of the black population in Soweto (a black township in Johannesburg) consumed indigenous medicine. Ellis [1986] found that 100% of a random sample of hospital patients in the Estcourt area (a typical rural population) used indigenous medicine. More recently, a study in peri-urban Bushbuckridge [Mander 1997a] estimated that 58% of the clinic patients used indigenous medicine. However, this estimate is known to be conservative due to the persecution of people suspected of witchcraft in the area, and consequently an unwillingness to admit to the use of indigenous medicine. Assuming that the characteristics of the Durban consumers are representative of the rest of the population in KwaZulu-Natal and South Africa, it is possible to provide a preliminary indication of provincial and national consumption patterns.

In KwaZulu-Natal the current population is estimated to be 8.7 million, with the black population comprising 83% of the total population or 7.2 million people [Central Statistical Services 1996]. Assuming that 84% of the black population use indigenous medicine in KwaZulu-Natal, the number of potential users of indigenous medicine could therefore be 6 million people. Based on an average frequency of uses of 3.34 visits/uses per year (this translates to 20 million uses a year), and assuming an average mass of 216.5g per use, then total quantity of medicinal plants used in KwaZulu-Natal could amount to 4 339 tonnes per year.

Similarly, at a national level there are 33 million blacks in South Africa [Central Statistical Services 1996] with an estimated 28 million (84%) users. Using the same visitation rate per year and the mass per use as above, the total national consumption could amount some 90 million uses per year with a mass of 19 500 tonnes of plant material.

Trends in the Use of Indigenous Medicine

The survey of consumer trends at western clinics and at indigenous healers indicated that a large section of the population has not changed their use patterns, with similar percentages indicating both an increase and a decrease in use. There is, however, an overall increase in the use of indigenous medicine. The following figures (4.3 to 4.8) and discussion outline the changes and reasons for change in use frequency cited by healers and clinic patients.

Of the indigenous healers interviewed, 90% believed that the demand for indigenous medicine would remain high. The shop traders shared the same perspective that trade would remain at high levels.

Discussions with market players and observations of the markets in southern Africa (South Africa, Mozambique, Swaziland, Lesotho and Namibia) have pointed to an increase in the demand for medicines [Scott pers. comm. 1996; Maseko pers. comm. 1997]. While the individual consumers surveyed may not have indicated a trend of increasing use, there is common consensus among traders and researchers that there has been an increase in the number of individuals consuming indigenous medicine. This may be the result of the following factors.

The population growth throughout the region is around 2.4% per year [Central Statistics 1996], leading to an increase in the potential number of people who may make use of indigenous medicine. Coupled with this increase, are the accelerated urbanization rates, which increase the competition for resources and services, especially where economic growth lags behind population growth. Indigenous medicines are actively used to reduce competition for employment, housing, and other social resources that are associated with increasing urban competition. Competition in urban areas may also contribute to psychological stress, and indigenous healing and healers are used extensively to treat stress-related illness.

The existence of poverty also fuels the demand for medicinal plants as households are forced to make use of affordable medicine. While consulting a healer may be more expensive than visiting a clinic, medicines bought directly from the street traders are much cheaper than any other forms of health care, and consequently are an important health care option.

Past government policies in South Africa discriminated against indigenous healing, giving it no positive recognition, and generally labelling the activities as primitive and even legislating against witchcraft practices (Suppression of Witchcraft Act). Perceptions are now changing from several quarters, with increasing official and societal recognition.

For example, medical aids in South Africa are increasingly accepting claims, by their members, for visits to indigenous healers [Daily News, 14 January 1997], and black people themselves are developing increasing pride in the culture of indigenous healing.

One of the greatest drivers of the demand for indigenous medicine are the widely held views by the black community that certain illnesses are 'cultural' sicknesses which can only be treated by indigenous medicine. In addition, AIDS is rapidly increasing within the South African community, and with little relief from western medicine, healers report that people with AIDS consult them. There are many healers who now advertise that they have a cure for AIDS, promoting the increased use of indigenous medicine.

Apart from the above, the remote geographical location of many large rural populations also promotes the use of indigenous medicine as it is usually the most accessible source of health care, and the most available health care, as western clinics are either far away or visit rural communities only periodically.

In view of the above factors, the local demand for indigenous medicine is unlikely to decline and will probably increase as the population increases.

Apart from growing local demand, there are also indications of a growing international demand. Neighbouring countries (particularly Namibia and Botswana) are increasing their demand for South African medicinal plants due to the greater diversity, reliability of supply and quantities available.

The dry environments have relatively low diversities of high value medicinal plants and in addition, the frequent and regular movement of people between southern African states results in considerable exchange in ideas and information, with foreign healers learning to use South African products. In addition, the declining ability of many countries to maintain the importation of synthetic drugs promotes the continued use of indigenous medicines8. International demand for local plant products is also likely to increase with the growing popularity of alternative medicines and natural products in developed countries [Lange 1997].

For example, discussions in 1996 with women from a remote area in western Mali, reported to the author that they did not bother travelling to the local clinic, in this case only a few kilometres away, as the only medicine which the clinic could provide was quinine and aspirin, and good alternatives for these drugs were more easily accessible in the bush close to where they lived.

Market Segments

Two major market segments, direct consumers and patients of indigenous healers, characterize the market for medicinal plants in KwaZulu-Natal. A third market segment is the pharmaceutical companies, but it is a minor market player at present.

The discussion will focus on the findings for Durban and will make projections for KwaZulu-Natal where appropriate.

Direct Consumers and the Indigenous Healers Patients

This market segment represents those consumers who purchase medicinal products for self-medication from street markets, shops, rural markets, and healers' practices and/or who purchase prescribed medicines from indigenous healers.

The survey was not able to interview direct consumers in the street markets and shops. Nevertheless, observations by the researcher and sales persons in the trade indicated that direct consumers and healers patients came from all socio-economic strata in the black community. Due to the survey limitations and the similarity between the two key market segments, the two market segments will be discussed together with differences indicated where they arise.

The survey was not able to make an assessment of the consumers from street markets and herb shops. The fast nature of transactions and the general pace of activity in the market did not make interviews possible. While sales persons were asked about consumer characteristics, the information obtained was not suitable for detailed analysis as it was based on genera! impressions.

With these limitations, no distinction will be made between the characteristics of the direct consumers (those buying for self-medication) and the healers patients. The report therefore uses the healers' patients as a proxy for the all the indigenous medicine consumers. Some information from the clinic patients is used where appropriate.

The Buyers—Who are They

The majority of users of indigenous African medicine in Durban, and for most of South Africa, are black. They represent a diverse group, with a wide range of social and economic characteristics. A number of figures are presented to indicate the range of consumer characteristics amongst the indigenous medicine users. At present there is limited use of indigenous medicinal plants by other population groups in South Africa.

This has important implications for the market potential as this study has shown that there is currently little or no differentiation in the products marketed while there are considerable differences in amongst the consumers. A survey of black clinic patients within Durban indicated that some 53% of the respondents visited indigenous healers in the last year and a further 31% may use indigenous medicine but did not visit healers in the previous year.

Assuming that the clinic sample represents the black population in Durban, then some 1.3 million people may make use of healers every year, with an additional 760 000 people using indigenous medicine without going to a healer, that is, probably purchasing products directly from traders. In total there are approximately 2 million indigenous medicine consumers in Durban, and at a provincial level, basing the estimates on the Durban statistics, there may be some 6 million consumers in KwaZulu-Natal, and 27 million in South Africa.

This assumption is valid as the socio-economic characteristics of the clinic patients are similar to that described by broader studies [Durban Metropolitan Government 1996] of the black population in the Durban region.

This is likely to be a conservative estimate given that some 50% of the population of Kwa Zulu-Natal is rural, and their frequency of indigenous medicine use is likely to be greater than the urban areas given the limited access to western bio-medical services in remote regions. This is likely to be the case for South Africa as a whole which has similar urban/rural population split.

It is interesting to note that the age-class distribution for the sample tends to mirror the age class distribution for the population as a whole [Durban Metropolitan Government 1996], indicating a broad spectrum of use. One expects that the use of healers would be greater in the older age classes of black society which are more traditional. This is not evident in the study.

The sample used is from the clinic patients as the indigenous healers sample was biased towards females on the days on which the survey was carried out. The survey was undertaken on days when large numbers of patients visited the healers focusing on the treatment of children. Hence the use of the Durban clinic sample as a more representative sample of the user population.

The education category distribution is similar to that recorded for the Durban region [Development Bank of Southern Africa 1994] and indicates that the sample selected is a relatively accurate representation of the Durban population. Past speculation has predicted that low education levels would be strongly correlated to the use of indigenous medicine. However, that over 60% of the consumers have at least some form of secondary education. Higher education levels do not appear to result in a reduction in the consumption of indigenous medicine.

One may also expect that certain religious affiliations are likely to influence the use of indigenous medicine, given the magic and ancestral worship associated with its use. The survey indicates that 96% of the consumers had a Christian affiliation.

A broad spectrum of occupational classes are represented in the survey sample. What is clear that the use of indigenous medicine is not confined to the lower earning occupations. This is confirmed in the figure illustrating the wide range of incomes earned by the consumers' household.

Confirms the wide range of the people that make use of indigenous medicine. The range includes, people from conservative traditional backgrounds (the traditional homesteads), relatively wealthy people (big houses), average people (4-roomed house), recent migrants to the city, new families, and people affected by political violence (the shacks), and single people and poor people (the hostels and rented rooms). The diversity in the types of homes which consumers live in provides an indication of the diversity of social and economic backgrounds of the users.

The above discussion shows that the majority of the black population in Durban is making use of indigenous medicine irrespective of religion, age, education and economic status. The direct consumers and the healers patients represent a large consumer population with considerable socio-economic diversity within the group. This has important implications for the market's development.

The Products Purchased

The products consumed are generally unrefined plant medicines with limited processing apart from grinding, chopping, and mixing. Raw plant material is frequently bought directly by consumers who may grind it themselves following instructions provided by the healers or traders. Plant products are taken as medicines in the following forms:

- infusions (chopped bark, leaves, seeds/fruit, stems, bulbs, rhizomes and roots steeped in water) which can be drunk, used as an emetic, or as an enema,
- concentrates (infusions where the water is boiled off concentrating the liquid) which are usually drunk,
- inhalants (plant parts or powdered parts burnt to produce smoke or boiled to produce steam) which is breathed in,
- powder (ground plant parts, usually tree bark, roots and leaves - usually burnt or raw) and which can be licked, used as snuff, bathed with, rubbed on to the skin, or implanted under the skin,
- poultices (fresh leaves of fleshy plants) applied to wounds, sores, and other skin ailments, and
- protective charms (live or dried plants) used for planting or scattering around the homestead.

The Quantities Demanded

The Durban survey estimated that some 1 500 tonnes of plant material was consumed per annum. The average mass per plant product bought in the street markets was 216.5g, and was 83g per product in the traders' shops. Observations in the street markets indicated that customers usually purchased one item in the street market, whereas in the traders' shops, customers tended to buy more than one product.

The shop traders reported that the customers buy on average 127.2g per visit or 1.5 items per visit. At the healers' practices, the customers purchased similar quantities to those traded by the shop traders but it was usually only one item per visit.

The healers' patients did not select the quantity purchased but were prescribed a quantity by the dispensing healer. In addition, the item purchased was usually a mixture of various plant species (which is generally confidential information).

The indigenous healers' patients or customers consumed some 61% of the market produce (936 tonnes), while the direct consumers bought 22% of the market produce from traders' shops (340 tonnes) and 17% from street traders (256 tonnes).

Patients refers to someone who is prescribed a medicine, whereas a customer is someone who purchases a product from the healer's stock without a consultation.

At a provincial level, the consumers demand approximately 4339 tonnes a year and nationally some 19 500 tonnes may be consumed.

The Timing of Purchases

Consumers visited indigenous healers on average 3.34 times a year and visited the markets towards end of the month when salaries were paid. Purchases of medicine were generally associated with doing other regular chores, such as travelling to work or going shopping. Visiting indigenous healers on the other hand usually involved a specific journey to the healer but with other activities attached.

This estimate is used to predict the visitation frequency of the entire user population (healers' patients and direct consumers) as the survey did not collect information on the frequency of indigenous medicine use (as opposed to the frequency of visiting a healer).

Observations of the market, and interviews indicated that there was greater buying activity on Mondays, Fridays and particularly Saturdays. Limited purchasing took place on Sundays. The time of purchases from shops and the street was normally between 10h00 and 16h00 during weekdays and on Saturdays between 10h00 and 14h00. Patients visiting healers tended to follow similar weekly patterns, but started earlier in the day. Patients arrived at the healers' practices between 07h30 until about 16h00 during weekdays, and up to 13h00 on Saturdays.

Consumption of certain species was generally constant throughout the year with a small peak in demand for chest-related medicines during the winter months and a small peak for homestead blessing products at the end of the year. The climate in Durban is tropical and consequently has a limited winter peak for chest-related medicines. In contrast, healers and traders in colder regions have reported substantial increases in the demand for Siphonochilus, Alepidea and Warburgia during the winter months (June, July and August in southern Africa).

The Reasons for Purchasing Indigenous Medicine

Consumers buy medicinal plants for a range of reasons as discussed. Various experts and authorities have speculated that the high demand for indigenous medicine by the black population in South Africa is due to low income, poor education opportunities, high costs of western medicine, and the lack of clinics.

An analysis of clearly the perspective that access to clinics and the cost of western medicines were not important considerations in determining the use of healers. On the other hand, strong sentiments were expressed regarding the desire for using indigenous health care systems and there was also strong support for the notion that clinics could not cure particular ailments. Respondents made it quite clear that the selection of healing systems depended on the type of illnesses or problems that they were experiencing. This indicates that in many cases, there are not western alternatives to the services provided by indigenous medicine. The survey obtained clear evidence indicating that black people in Durban make use of both the western and indigenous health care systems.

The survey showed little or no evidence of any positive relationship between low education and income levels and a higher frequency of visits to healers. On the contrary, the healers' customers reported that they would continue to use indigenous medicine even if it became more expensive. In view of the above attitudes, indigenous medicine could be classified as a basic consumer good, such food, clothing, western medicine and housing.

In fact, consumers reported that indigenous healers were already more expensive than western medicine.

While the cost of consulting healers may be relatively expensive when compared to clinics, it is important to note that self-medication, using indigenous medicines, was the cheapest form of health care available.

This is especially relevant to rural communities. Unlike the cities, where clinics may be relatively easily accessible, rural areas have few clinics and consequently rural communities have little or no choice in the health care systems used. Indigenous health care being the only accessible system for a large proportion of the people given the distances to travel to clinics and the costs of transport.

The Purchasing Power of Consumers

The purchasing power of consumers can be considered from the perspective of the household or from the individual consumer. Household income is discussed above and shows that there is a wide range in purchasing power, with the majority earning relatively low incomes and therefore individuals having limited purchasing power. However, purchasing power is unlikely to be a significant limitation when buying products for self-medication as this could amount to US$ 1.50 (R 6.70) (the cost of two beers) per year for an individual consumer or US$ 9 (R 40) per household per year.

Using an average of 3.34 purchases of indigenous medicine per year. The average household size for healers' customers is six people. On the other hand, consulting an indigenous healer is likely to represent a significant cost as the average consulting fee (US$ 8.2 or R37) and associated prescribed medicine is some 18 times more expensive than the average cost of medicine bought at the markets for self-medication, and could cost an individual consumer around US$ 28 (R 124) per year. The costs to a household (of six people) could amount to US$ 168 (R 744) per year. The average monthly income earned by households visiting healers is between US$ 178 (R 800) to US$ 333 (R 1 500) per month. Consequently, the average household could spend between one half a month's income (4.2% of annual income) and one month's income (8.3% of annual income) on indigenous medicine per year.

The traders (street, shops and healers) indicated that at least 75% of their customers accepted the price charged, with the remainder bargaining for a lower price. This may indicate that for most consumers, the prices charged are affordable.

While the purchasing power of a large segment of the population is limited due to low income and relatively high costs (in prescribed medicine), it is unlikely to be limited for purchasing products for self-medication. In addition, the consumers indicated that their demand would remain the same irrespective of high prices.

'High prices' are probably relative to past experiences of price increases which, if considered by percentage, are considerable (for example, 50% increases in prices are common). However, given the real amount of money involved (US$ 0.11 or R 0.50 increase per product purchased) and the infrequent use of indigenous medicine, it does not constitute a significant cost.

At a community level, the purchasing power of the user group in Durban is significant and amounts to some US$ 1.02 billion (R 4.6

billion) per year. The survey has shown that the demand for indigenous medicine is relatively inelastic, that is, the consumers consider indigenous medicine as a basic consumer good which needs to be purchased to maintain their well-being. Consequently, there is a high degree of certainty that considerable sums of money will be spent on indigenous medicine on an annual basis in Durban. Using the estimated annual expenditure per household on indigenous medicine (between 4.2% and 8.3% of annual household income), then one can estimate that between US$ 42 million (R 190 million) and US$ 84 million (R 380 million) could be spent per annum in Durban.

This estimate is based on a user population of 2 million, with an average monthly income of R1 150 (US$ 256) and an average household size of six people [Development Bank of Southern Africa, 1994]. Using the quantities of plants traded and the average price of raw products bought (R 14/kg), it is possible to estimate the value of expenditure on raw plant products.

These values are the trade in raw products and exclude any value-added through processing or prescription.

In Durban some 1 500 tonnes were traded and this would have generated an expenditure of US$ 4.7 million (R 21 million) per annum.

In KwaZulu-Natal, the trade of 4 300 tonnes would have generated an expenditure of some US$ 13.3 million (R 60 million) per annum. This is approximately one-third of the value of the annual maize harvest in KwaZulu-Natal.

At a national level, a trade of 19 500 tonnes would generate an expenditure of US$ 61 million (R 273 million) per annum (assuming the average price of Durban products).

Who are the Buyers: Several pharmaceutical companies are currently developing products for the African market in South Africa. There are reports that some companies are buying raw materials from plant harvesters in rural areas. One pharmaceutical company, which has a major share of the domestic market for western bio-medical products, is currently establishing partnerships to cultivate plants for their production process. Very little information is available regarding the pharmaceutical companies given the high degree of competition and secrecy that exists.

It is known that one of the companies is currently focusing on producing products for the upper-end of the consumer market. The large pharmaceutical company, is also focusing on products which would be registered by the Medicines Control Council, which has high standards matching the conventions in Europe and North America.

Quantities Demanded: The quantity of plants demanded is unknown at present, and is likely to remain so for the short term until the practices become more established and transparent.

The Products Purchased: A small number of plants are being focused on due to the high costs of pharmacological screening and testing. Some 10 to 20 species are probably the focus of current formal commercial activities. The focus is also likely to be on species, which are relatively easily cultivated to ensure that commercial production can be sustained.

Commercial activities undertaken by conventional businesses (in the western sense) rather than informal markets which dominate the current trade in indigenous medicine.

The Timing of Purchases: Whilst the products are still being developed and consumer preferences have not been tested in the market place, the timing of purchases is likely to be unknown.

The Reasons for Purchases: Pharmaceutical companies purchase material to process and trade to consumer outlets. The business opportunities associated with the large numbers of indigenous medicine consumers have, until recently, been largely ignored by the pharmaceutical industry. Recent research into the size of the indigenous medicine market has fuelled considerable interest by local companies.

The Purchasing Power: The purchasing power of large pharmaceutical industries is considerable relative to other market players in the trade. The access of large corporations to finance is far greater than the monies available to the operators in the current indigenous medicine trade. However, despite the greater purchasing power, established pharmaceutical companies are unlikely to purchase large volumes of material from existing traders, due to the unreliability of supply and current legislation.

The Supply of Medicinal Plants in KwaZulu-Natal

This study has established that the large volume of indigenous medicinal plants, which is traded in the southern African markets, is harvested from wild populations, with insignificant cultivation taking place. These plants are harvested from within South Africa and from neighbouring countries such as Swaziland, Lesotho, Mozambique and Namibia. The supply of high value plants has declined leading to concerns within the industry and conservation agencies.

Sources of Supply

Medicinal plants supplied to the markets in KwaZulu-Natal are harvested from a wide range of habitats in the region, including coastal forests, coastal grasslands, mangroves, swamp forests, grasslands, woodlands, riverine forest, montane forest, wetlands, and sub-alpine grasslands. These habitats are located on communal lands, commercial farms, forestry estates, protected areas, and in neighbouring countries.

Domestic Sources: The majority of the plants are harvested from wild populations on communal lands. Popular species, which are no longer available in these communal areas, may be harvested on forestry estates (exotic plantations with natural areas), commercial farms, and protected areas. Harvesting takes place with or without the consent of the landowners or local authority. There is limited commercial cultivation of medicinal species for local markets, although several organizations, such as Silverglen Medicinal Nursery, Institute of Natural Resources, Mondi Forests (a timber company) and several conservation agencies are promoting the cultivation of medicinal plants by farmers and other commercial enterprises.

Supply from wild populations in communal areas: The main source of plants for markets in KwaZulu-Natal is from communal or tribal lands where there is generally easy access to forest and grassland resources. In South Africa, there has been an erosion of tribal authority, and little or no state support for developing management systems for communal natural resources. Consequently, there are limited skills and motivation to promote natural resource management in these areas. This has led to a situation where rural communities take little or no responsibility for the state of their natural resources, allowing individuals in the community or outsiders to exploit medicinal plant populations with little or no control [Mander 1997a].

The communal areas closest to the urban and rural markets have been the primary focus of harvesting activities. However, as high value or popular plants have been depleted in these areas, harvesting of popular and scarce species takes place in more and more distant areas.

The location of the main harvesting areas has increased in distance from Durban, with an average increase in travel time6 of 45% (or 2 hours) over 8 years, between the source and market. The change in travel time ranges between an increase of 1 hour to 4 hours. For example, Eucomis autumnalis, a formerly common grassland bulb which was harvested largely within two hours travel, is now being extensively harvested in areas of 4 to 5 hours away. Furthermore, other barriers to trade, such as international borders and the associated trade controls, are increasingly being encountered and overcome in order to meet the demand for plants. Six out of the nine most popular medicinal species being studied (66%) are now being harvested in either Swaziland and/or Mozambique, and transported to the Durban markets for sale. Travel time has been used to give an indication of changes in the distance between the markets and the source as it represents a more realistic barrier in the market than distance.

The declining size of bulbs and thickness of bark also illustrate the scarcity of plants. Indigenous healers have reported that they are increasingly using smaller sizes of plants, indicating that popular plants are getting less and less opportunity to grow and to reproduce. Between 25% and 39% of the healers indicated that, for the nine popular plants, they were now using smaller sizes than before. Between 8% and 15% of the healers indicated that they were now using different plant parts.

The general lack of willingness and/or ability to enforce property rights associated with plants in the region, combined with the high levels of poverty in rural areas, exacerbates the destructive methods of harvesting. Rural women are desperate to generate a cash income, and will therefore harvest large numbers of plants for sale to the markets. Sustainable harvesting is not practised, as harvesters have no guarantees that they will benefit from any wise harvesting of species. The above factors result in a highly destructive approach to harvesting with each high value plant population becoming depleted, leading to a new focus on either a different species or on a further area which may still have high value plants.

A further destructive condition, which impacts on plant populations, is the fact that most women harvesters concentrate their activities in one area. They are unlikely to shift to new areas to look for a particular species due to travel costs, and will focus on different species to harvest. This means that a population of popular species will not be given a chance to regenerate as any individual popular plant seen while in the course of gathering other plants, will be collected. Consequently, popular species face localized extinctions in areas where harvesters regularly gather medicinal plants for the market.

As consequence of the above approaches to harvesting, the supply of indigenous plants from communal areas is declining, leading to a shortage of popular plants in the market. The local extinctions and concomitant increasing distance to harvesting locations, and the declining size of the plants utilized by healers, are clear indicators that the current supply of plants from communal areas is not sustainable. This has resulted in an increase in imports and an increasing interest in cultivation.

Supply from Commercial Farms: There are several commercial operations initiating the cultivation of indigenous plants to supply markets. Production is focused on supplying high value plants to pharmaceutical companies either within South Africa or in European markets. Production for the South African market is believed to be limited to a few species that can be registered with the Medicines Control Council. In KwaZulu-Natal there are low supplies of plants from commercial farms.

Due to the infancy of large-scale commercial ventures in the indigenous medicine markets and the lack products on the market, there is considerable secrecy in the industry. Little information will be available until products are openly traded.

The level of harvesting from undeveloped areas on commercial farms is unknown. The localities identified by the plant gatherers indicate that there is limited harvesting on commercial land. There is however, some illegal harvesting from commercial farms or estates. There has been an increasing awareness amongst landowners regarding the value of medicinal plants, and it is unlikely that any large-scale harvesting takes place on commercial farms. Harvesting plants for home consumption by farm labourers is likely to be the largest form of use on these properties.

Imports: A number of plant species are imported into South Africa (including KwaZulu-Natal) for the medicinal plant trade. The imports take place on an informal basis with no documentation or customs authorization. There is considerable labour migration between Lesotho, Swaziland, Mozambique and South Africa, and this promotes a large informal movement of plant material. Rural families from neighbouring countries frequently have members of their households or community working within South Africa, particularly in large cities, thereby providing remote locations with market information and transport mechanisms. The material is usually traded in its raw form and can be either sold on street markets, in shops, by healers or by wholesalers. Popular species, now becoming increasingly difficult to obtain in South Africa, are the main focus of trade. Discussions in the Durban market have indicated that Haworthia limifolia, Boweia volubilis and Warburgia salutaris are imported in large quantities from Swaziland and Mozambique. These two countries are now recognized as the main sources of supply for these species. Two other species in great demand, Alepidea amatymbica and Eucomis autumnalis, are imported from Lesotho but in small quantities.

The different sources of imports can be ascribed to the distribution of species. All of these species are still found in South Africa but the declining local supply and increasing in-accessibility has led to the importation of these species from countries where they are still common and/or easily accessible.

The quantities of plants imported into South Africa are unknown. The informal nature of the plant trade and the illegal nature of imports make it difficult to make any quantitative estimates of the quantities imported into South Africa.

For one species, Warburgia salutaris, it may be possible to provide some idea of what is being imported due to the localized extinction that has occurred in parts of South Africa. The market survey estimated that some 17 tonnes of Warburgia salutaris were traded in Durban in 1996 and Williams pers. comm. 1997 suggests that between 10 and 20 tonnes were traded in Gauteng (formerly the Johannesberg area) in 1995.

As the Durban and Gauteng markets are the major trading conduits, it is likely that between 30 to 40 tonnes of Warburgia salutaris are imported annually from Mozambique and Swaziland. Some of the other species, such as Boweia volubilis, Siphonochilus aethiopicus, Curtisia dentata, Eucomis autumnalis, and Haworthia limifolia are also imported to some degree but the quantities are unknown. For example, 33% of the traders selling Boweia volubilis (or nine traders), indicated that they obtained their plants in Mozambique.

Bibliography

Arguello D. S.: *Production, Post-harvest Management and Exportation of Tropical Fruits in Costa Rica*, Wageningen, Technical Centre for Agricultural and Rural Cooperation, 1992.

Arora, Chandralekha : *Advances in Horticulture Science Research*, Oxford Book Company, Delhi, 2009.

Bhattacharjee, Supriya Kumar : *Amenity Horticulture, Biotechnology and Post-Harvest Technology*, Pointer, Delhi, 2006.

Chadha K. L. and Pareek O. P.: *Advances in Horticulture: Fruit Crops,* New Delhi, Malhotra Publishing House, 1993.

Crucefix D.: *Avocado Variety Selection for Export Development,* Roseau, CARDI, 1996.

Currah L. and Proctor F. J.: *Onions in Tropical Regions*, Kent, Natural Resources Institute, 1990.

Dhatt A. S. and Singh Z.: *Propagation and Rootstocks of Citrus*, New Delhi, Malhotra Publishers, 1993.

Doijode S. D.: *Seed Germination in Fruits*, New Delhi, Malhotra Publishers, 1993.

Featherly H. I.: *Taxonomic Terminology of the Higher Plants*, USA, Iowa State College Press, 1954.

Glowinski Louis: *The Complete Book of Fruit Growing in Australia*, Australia, Lothian Publishing Company Pty. Ltd., 1991.

Gunjate R. T. and Tawde A. B.: *Propagation and Rootstocks of Mango*, New Delhi, Malhotra, 1993.

Harlan, Jack R.: *The Living Fields: Our Agricultural Heritage*, Cambridge: Cambridge University Press, 1998.

Herklots G. A. C.: *Vegetables in South East Asia*, London, George Allen & Unwin Ltd., 1972.

Jeffers P.: *Evaluation of Four Onion Varieties in Montserrat*, Plymouth, CARDI, 1992.

Jerram, M.R.K. : *A Text-Book on Forest Management : In Agriculture, Horticulture and Forestry*, Asiatic Pub, Delhi, 2006.

Jugale, V.B. : *Horticulture Economy of Maharashtra*, Shruti, Delhi, 2004.

Malins A.: *Postharvest Handling of Pineapple and Mango*, Port of Spain, Trinidad and Tobago, 1992.

Matsuoka H.: *Cultivation of Panicum Genetic Resources for Evaluation of Characteristics*, Tokyo, JICA, 1997.

Mitra S.: *Postharvest Physiology and Storage of Tropical and Subtropical Fruits*, Oxon, CABI, 1997.

Morton Julia F.: *Fruits of Warm Climates*, Miami, Julia F. Morton Publisher, 1987.

Oldham P.: *Cost of Production of Major Tree Crops in Dominica*, Roseau, Ministry of Agriculture, 1991.

Peter, K.V. : *Basics of Horticulture*, New India, Delhi, 2009.

Pilgrim R.: *Post Harvest Handling of Minor Exotics*, St. George's, Grenada, 1996.

Prasad, Surendra : *Agro's Dictionary of Horticulture*, Agrobios, Delhi, 2003.

Sharma, V.K. : *Advances in Horticulture : Strategies, Production, Plant Protection and Value Addition*, Deep and Deep, Delhi, 2004.

Singh H. P. and Chadha K. L.: *Genetic Resources of Citrus*, New Delhi, Malhotra Publishing House, 1993.

Singh, D.K. : *Elements of Horticulture*, Agrotech, Delhi, 2005.

Singh, H.P. : *Challenges of Climate Change - Indian Horticulture*, Westville Publishing House, Delhi, 2010.

Singh, S.P. : *Advances in Horticulture and Forestry*, Scientific, Delhi, 2001.

Thavarasook C.: *Proceedings of the Mungbean Meeting 90, Chiang Mai, Thailand,* Bangkok, Tropical Agricultural Research Centre, 1991.

Whitwell A.: *Dominica Orchard Crop Management and Research Project: Pest and Disease Management,* United Kingdom, Natural Resources Institute, 1991.

Index

❑❑❑